SELECTED CONTRIBUTIONS TO GAME THEORY

World Scientific Series in Economic Theory

(ISSN: 2251-2071)

Series Editor: Eric Maskin *(Harvard University, USA)*

Published

Vol. 11 *Selected Contributions to Game Theory*
by Robert J Aumann

Vol. 10 *Interactive Epistemology*
by Robert J Aumann

Vol. 9 *Bayesian Persuasion*
by Matthew Gentzkow and Emir Kamenica

Vol. 8 *Learning and Intertemporal Incentives*
by Dirk Bergemann and Juuso Välimäki

Vol. 7 *Models of Bounded Rationality and Mechanism Design*
by Jacob Glazer and Ariel Rubinstein

Vol. 6 *Uncertainty within Economic Models*
by Lars Peter Hansen and Thomas J Sargent

Vol. 5 *The Language of Game Theory: Putting Epistemics into the Mathematics of Games*
by Adam Brandenburger

Vol. 4 *Simple Adaptive Strategies: From Regret-Matching to Uncoupled Dynamics*
by Sergiu Hart and Andreu Mas-Colell

Vol. 3 *Case-Based Predictions: An Axiomatic Approach to Prediction, Classification and Statistical Learning*
by Itzhak Gilboa and David Schmeidler

Forthcoming

Decision Theory
Wolfgang Pesendorfer *(Princeton University, USA)* &
Faruk Gul *(Princeton University, USA)*

Leverage and Default
John Geanakoplos *(Yale University, USA)*

Leverage Cycle, Equilibrium and Default
Vol. 2: Collateral Equilibrium and Default
John Geanakoplos *(Yale University, USA)*

More information on this series can also be found at http://www.worldscientific.com/series/wsset

World Scientific Series in Economic Theory – Vol. 11

SELECTED CONTRIBUTIONS TO GAME THEORY

Robert J. Aumann
The Hebrew University of Jerusalem, Israel

World Scientific

NEW JERSEY • LONDON • SINGAPORE • BEIJING • SHANGHAI • TAIPEI • CHENNAI

Published by

World Scientific Publishing Co. Pte. Ltd.
5 Toh Tuck Link, Singapore 596224
USA office: 27 Warren Street, Suite 401-402, Hackensack, NJ 07601
UK office: 57 Shelton Street, Covent Garden, London WC2H 9HE

Library of Congress Cataloging-in-Publication Data
Names: Aumann, Robert J., author.
Title: Selected contributions to game theory / Robert J. Aumann,
The Hebrew University of Jerusalem, Israel.
Description: New Jersey : World Scientific, [2025] | Series: World Scientific series in
economic theory, 2251-2071 ; vol. 11 | Includes bibliographical references.
Identifiers: LCCN 2020022334 | ISBN 9789811221064 (hardcover) |
ISBN 9789811222610 (ebook) | ISBN 9789811222627 (ebook other)
Subjects: LCSH: Game theory. | Economics--Mathematical models.
Classification: LCC HB144 .A947 2025 | DDC 519.3--dc23
LC record available at https://lccn.loc.gov/2020022334

British Library Cataloguing-in-Publication Data
A catalogue record for this book is available from the British Library.

For any available supplementary material, please visit
https://www.worldscientific.com/worldscibooks/10.1142/11851#t=suppl

Desk Editor: Lum Pui Yee

Typeset by Stallion Press
Email: enquiries@stallionpress.com

Contents

Preface xi

Introduction xiii

About the Author xxv

Chapter 1. Acceptable Points in General Cooperative n-Person Games 1

1.1 Introduction . . . 1
1.2 Notation . . . 2
1.3 General Cooperative Games . . . 5
1.4 Acceptable Points . . . 7
1.5 Two-Person, Zero-Sum Games . . . 8
1.6 Strategies and Payoffs in the Supergame . . . 10
1.7 Strong Equilibrium Points . . . 16
1.8 Application of Approachability and Excludability Theory . . . 17
1.9 The Main Theorem: First Half . . . 20
1.10 The Main Theorem: Second Half . . . 22
1.11 Existence of Acceptable Points . . . 37
1.12 The Expected Payoff . . . 38
1.13 Enforceability of Agreements . . . 40
References . . . 42

Chapter 2. A Definition of Subjective Probability **43**

Joint with F. J. Anscombe

2.1 Introduction 43
2.2 Lotteries 44
2.3 Assumptions 45
2.4 Existence of Subjective Probabilities 47
2.5 Comparison with the Literature 50
Acknowledgements 52
References 52

Chapter 3. Markets with a Continuum of Traders **55**

3.1 Introduction 55
3.2 The Mathematical Model and the Main Theorem 60
3.3 Discussion of the Model 62
3.4 Proof of the Main Theorem 63
3.5 Discussion of the Literature 67
References 70

Chapter 4. Subjectivity and Correlation in Randomized Strategies **73**

4.1 Introduction 73
4.2 Examples 75
4.3 The Formal Model 82
4.4 Preliminaries and Generalities 86
4.5 Equilibrium Points in Two-Person Games 89
4.6 Two-Person Zero-Sum Games 90
4.7 Proof of the Propositions of Section 4.4 93
4.8 *A Posteriori* Equilibria 97
4.9 Discussion 99
Acknowledgment 112
References 113

Chapter 5. Values of Markets with a Continuum of Traders **115**

5.1 Introduction 115
5.2 The Value in Finite Games 118

5.3 The Value in Continuous Games 119
5.4 Differentiability, Smoothness, and Uniform Smoothness . 121
5.5 Description of Model and Statement of Results 123
5.6 Interpretation 126
5.7 An Example 130
5.8 An Unrigorous Demonstration of the Results 132
5.9 Discussion of Smoothness and Some Counter-Examples . 137
5.10 Discussion of the Uniformity Conditions 143
5.11 The Theory in a Cardinal Context 144
5.12 Discussion of the Literature 149
5.13 Proof of Proposition 5.1 151
5.14 Toward the Proof of Proposition 5.2 152
5.15 Proof of the Value Equivalence Theorem 156
Appendix: Marginal and "True" Contributions 160
References 167

Chapter 6. Agreeing to Disagree **169**

References 173

Chapter 7. Correlated Equilibrium as an Expression of Bayesian Rationality **175**

7.1 Introduction 176
7.2 Correlated Equilibrium 178
7.3 Bayesian Rationality in Games 183
7.4 Discussion 185
7.5 The Common Prior Assumption 192
7.6 Randomness as an Expression of Ignorance 196
7.7 The Literature 198
References 199

Chapter 8. Long-Term Competition — A Game-Theoretic Analysis **201**

Joint with L. S. Shapley

8.1 Introduction 201
8.2 Repeated Game Models 203
8.3 Perfect Equilibrium Points 207

8.4 Characterization of PEPs 209
8.5 Discounted Payoffs in Repeated Games: Discussion of an Example 212
References 218

Chapter 9. Game-Theoretic Analysis of a Bankruptcy Problem from the Talmud 219

Joint with M. Maschler

9.1 Introduction 219
9.2 The Contested Garment 222
9.3 Consistency 224
9.4 Self-Duality and Constrained Equal Division 228
9.5 Coalition Formation and the Jerusalem Talmud 232
9.6 The Nucleolus and the Kernel 235
9.7 Historical Notes 240
Acknowledgments 241
References 241

Chapter 10. Game Theoretic Aspects of Gradual Disarmament 243

Joint with M. Maschler

10.1 Introduction 243
10.2 A Case When a Player Cannot Benefit in the Long Run from His Information 246
10.3 A Case When a Player Should Disclose His Information 252
10.4 A Case When a Player Should Partially Disclose His Information 254
10.5 The Main Theorem 263
10.6 The Optimal Strategy of Player 2 276
10.7 Conclusions 279
10.8 Where Do We Go From Here? 280
Postscripts to Chapter Ten 283
a. Martingales 283
b. Convexity and Concavity as Monotonicity in Information 284

c. The Error Term 285
d. Optimal Strategies of the Uninformed Player 294
e. Monotonicity of v_n and the Recursive Structure of Γ_n 302
References 304

Index 307

Preface

We are grateful to many individuals, without whom this volume would not have appeared. Most especially, to Eric Maskin, the editor of the World Scientific Series in Economic Theory. First, for asking me to contribute a selection of my publications to this series. Even more important, for choosing the publications to be included in the volume.

When Eric originally suggested this volume, he asked me to make the selection. But one's publications are a little like one's children: One does not choose between them. The Talmud says, "A baker does not testify as to his own dough." Similarly, a scientist does not have the perspective to judge which of his discoveries are the most significant.

So, I asked him to make the selection. And he did. With great aplomb.

Introduction

The ten papers comprising this volume appear in it in the order in which they were published. Three of the ten chapters (1, 8, 10) deal with repeated games, two (4, 7) with equilibrium play, two (3, 5) with the law of supply and demand, and one each with the foundations of probability (2), common knowledge (6), and fair division (9). In discussing the papers here, we adhere to this grouping.

Game Theory has two major (interrelated) branches: the *strategic* (a.k.a. *non-cooperative*) theory, which is about how to play, what moves to make, and the *cooperative*, which is about sharing the proceeds. Technically, cooperative games are endowed with an exogenous mechanism for enforcing agreements, such as a court system; the question then is, what agreements are called for? In strategic games, no such exogenous mechanism is available, as is the case, for example, in international relations. In this collection, Chapters 4, 7, and 10 deal with the strategic theory, 3, 5, and 9 with the cooperative, 1 and 8 with bridging between the two, and 2 and 6 with foundational issues.[1]

Repeated Games

Complete Information

Many game situations have an ongoing element. Often, when people interact, they have interacted in the past, and expect to do so

[1]For publications mentioned in this introduction, the reader will find precise bibliographic information in the corresponding chapter.

again in the future. Informally, this ongoing element constitutes an endogenous mechanism for enforcing agreements; people adhere to agreements today because there is a tomorrow. The theory of repeated games provides a formal expression of this informal insight. Inter alia, the theory "predicts" phenomena like cooperation, altruism, secrecy, trust, punishment, and revenge, and deals with the specific forms that these phenomena may take.

To fix ideas, consider a single n-person game G, repeated again and again, indefinitely often. "Indefinitely" is modeled by "infinitely"; with the former, as with the latter, at each stage of the repetition, each player must take into account how the others will react at future stages. The repetition is itself a game, denoted by G^{∞}; its payoff is taken to be an average[2] of the stage payoffs. In particular, the attainable payoffs in G^{∞} are the same as in G.

The starting point for Chapters 1 and 8 is the so-called

Folk Theorem: A n-tuple h of payoffs to the players is associated with a Nash equilibium of G^{∞} if and only if it is feasible[3] and individually rational[4] in G.

Conceptually, feasible individually rational payoff vectors in G represent possible agreements, i.e., possible outcomes when G is viewed cooperatively. Clearly an agreement must be feasible; it must also be individually rational, as no rational player would agree to accept less than what he can assure himself. On the other side of the equation, Nash equilibria represent the strategic, "non-cooperative" approach.[5] Thus, the folk theorem formally identifies cooperation in the one-shot game with strategic behavior in the ongoing game — precisely the intuition laid out above.

---

[2] "An" average rather than "the" average because an infinite sequence of numbers does not have a well-defined average. The difficulty can be circumvented in various ways, which we do not discuss here.

[3] I.e., a convex combination of payoff vectors to n-tuples of pure strategies in G.

[4] Roughly, this means that no player i can assure himself of more than h_i. When there are more than two players, the precise definition is that no coalition of players can, each using mixed strategies, prevent i from getting at h_i.

[5] Indeed, "Non-cooperative Games" is the title of the groundbreaking 1953 paper in which Nash introduced his equilibrium.

Chapters 1 and 8 extend the Folk Theorem in two directions. In Chapter 1, individual rationality is replaced by group rationality, which means that no group of players can assure that each player in the group gets a payoff larger than that provided by the agreement. The set of feasible group rational payoff n-tuples is called the *core* of a cooperative game G. Correspondingly, Nash equilibrium is replaced by *strong* equilibrium, defined as a strategy n-tuple s such that no group of players can increase the payoff to each of its members by changing the strategies of the group's members, while each player not in the group plays as in s (Nash equilibrium applies this condition to single players only, not groups). Then, we have (roughly) the

Strong Folk Theorem: An n-tuple h of payoffs to the players is associated with a strong equilibium of G^∞ if and only if it is in G's core.

Chapter 8, coauthored with Lloyd Shapley, extends the Folk Theorem by replacing Nash equilibria in G^∞ with (subgame) perfect equilibria, namely n-tuples of strategies that remain in equilibrium when restricted to any tail of the repetition (i.e., for all k, and all sequences of moves up to stage k, the game consisting of stages $k, k+1, k+2, \ldots$). Explicitly, we have the

Perfect Folk Theorem: An n-tuple h of payoffs to the players is associated with a perfect equilibium of G^∞ if and only if it is feasible and individually rational in G.

This is important because the standard proof of the Folk Theorem relies on equilibria that are *not* perfect — specifically, on threats that it would be irrational to carry out, so are not credible.

Chapter 1, my first Game Theory publication, is perhaps a little technical. Chapter 8 was written in my prime and, moreover, has Lloyd Shapley's master touch; it is much more readable.

Incomplete Information

In Chapters 1 and 8, all players know all about the game being repeatedly played; the only thing they do not know is how the others will play. Chapter 10 relaxes this condition.

Let G_1 and G_0 be two-person zero-sum games with equally sized payoff matrices. Nature chooses one of them with probability $\frac{1}{2}$ and informs one player, P_I, but not the other, P_U, of its choice.[6] They then play the chosen game repeatedly. After each stage, each player is informed of the other's strategy at that stage, but not of his[7] own payoff,[8] which is deposited in his bank account.[9] The payoffs to the repeated game G^∞ are long-term averages, as in the case of complete information.

Surprisingly, to calculate the value of G^∞, one *must* consider a more general situation, where G_1 and G_0 are chosen with arbitrary probabilities p and $1-p$. Let $u(p)$ be the value (to P_I) of the one-shot game $pG_1 + (1-p)G_0$. The value of the repeated game is then $(\text{cav}u)(\frac{1}{2})$, where $(\text{cav}u)$ is the *concavification* of u, i.e., the pointwise smallest concave function of p that is $\geqq u(p)$ for all probabilities p.

In general, let $G^\infty(p)$ be the repeated game that is like G^∞, except that the stage games G_1 and G_0 are chosen with probabilities p and $1-p$ rather than $\frac{1}{2} - \frac{1}{2}$. Denoting the value to P_I by val, we then have the

Concavification Theorem: $val(G^\infty(p)) = (\text{cav}u)(p)$.

As explained in the body of the chapter, this says, roughly, that *in the long run, secrets can be kept only to the extent that they're not used.*

Chapter 10 is the first chapter of a 1995 book coauthored with my lifelong research partner Mike Maschler, which studies repeated

[6]The subscripts I and U stand for informed and uninformed.

[7]We adhere to the old convention that, in general, masculine pronouns are used for indeterminate gender.

[8]The informed — but not the uninformed — player can deduce the stage payoff from his information.

[9]The "bank account" is a euphemism meant to express the fact that the payoff to the uninformed player may depend on those very factors of which the uninformed player is uninformed. Thus, in the arms control context that originally motivated this study (see below), the payoff to the US for an agreement to destroy a given number of nuclear bombs depends not on the number that the Soviet Union destroys but on the number remaining, of which the US is uncertain.

games with *incomplete* information — ongoing interactive situations in which participants need not possess all the relevant information; if possible, they must glean it from the actions of the others. Though published only some 30 years later, the research on which the book reports was done in the period 1966–1968. This was the height of the 1946–1989 cold war between the United States (US) and the Soviet Union (SU), who were engaged in an intensive nuclear arms race. Both wanted to slow — perhaps reverse — the alarming increase in these dangerous, expensive weapons systems, but neither could afford to do so without being reasonably sure that the other also would. It was in this atmosphere that the two agreed on a series of conferences, to be held in Geneva, designed to craft agreements to achieve such a slowdown or reversal, in particular, to destroy a specified number of nuclear weapons on each side.

Mike and I were part of a group of specialists[10] retained by the US government to study the decision-theoretic aspects of this situation. We modeled the repeated Geneva conferences as repeated games with incomplete information. Here, unlike in Chapter 10, the information was incomplete on both sides — each knew only the number of weapons it had, not the number of weapons the other side had. Moreover, the game was not zero-sum. But Mike and I decided to tackle the difficulties one by one. So we started with the simplest case — two-person zero-sum repeated games with incomplete information on one side only. Later chapters of the 1995 book extend the results of the opening chapter to two-sided incomplete information, to general sum games, and in other directions. But the insights of the opening chapter underlie the entire research.

Common Knowledge: The Agreement Theorem

A group of people is said to have *common knowledge* (CK) of an event if all in the group know that the event obtains,[11] all know that

[10]Which included luminaries such as Gerard Debreu, John Harsanyi, and Reinhard Selten — all eventually to be awarded Nobel Prizes.

[11]I.e., happened, is happening, or will happen.

all know it, all know the foregoing, all know the foregoing, and so on ad infinitum. CK was first formally defined by the philosopher David Lewis, who in his 1969 book *Convention*[12] pointed out that the effectiveness of many conventions — like driving on the right — depends on their being CK. A good bit of Game Theory implicitly assumes that the rules of the game being played, and the payoffs, are CK among the players.

Chapter 6 formally states and proves what has come to be known as the "Agreement Theorem": Two people who base their probability assessments on their information only cannot "agree to disagree"; i.e., they cannot have common knowledge of differing probability assessments. This is so even though one's information is entirely different from the other's. Formally, if it is CK between Ann and Bob that to an event E, Ann assigns probability p_A and Bob assigns probability p_B, then $p_A = p_B$.

The key assumption here is that Ann and Bob base their probability assessments on their information only, not, say, on their preferences. For example, if Ann likes Charlie and Bob does not know him, then for the event E that Charlie committed murder, they might have CK of probabilities p_A and p_B with $p_A < p_B$ because Ann does not like to think that Charlie could do such a thing. In the formal development of Chapter 6, this key assumption is embodied in the assumption of "common priors" — that if they had no information at all, Ann and Bob would assign the same probability to E.

Mathematically, the proof of the theorem is neither difficult nor deep, but it is not obvious. I personally have attempted to explain it in words, without success.

Intuitively, Ann's information is somehow summarized in p_A and Bob's in p_B. Thus, Ann's total information consists of the information that she gets directly, plus p_B, which summarizes Bob's information; so in a sense, it represents the information of both. Similarly, p_B represents the information of both; so, since the probabilities are based on information only, they must be the same. Of course, that is nowhere near a proof.

[12]Of which I was unaware when first publishing Chapter 6.

Correlated Equilibrium

Chapter 4 describes extensions in two different directions of the fundamental notion of strategic equilibrium defined by Nash. One is *subjective* equilibrium — replacing the objective randomizations that define the mixed strategies in Nash equilibrium by subjective randomizations. In matching pennies, for example, the players could peg their choices on the outcomes of two horse races, rather than on spins of a roulette wheel. The other extension is relaxing the requirement that the randomizations defining the players' mixed strategies be independent. For example, in the game known as the "battle of the sexes," the players could use the same coin toss to choose between the boxing match and the ballet. Both extensions were seminal, in diametrically opposed senses: The first because it led nowhere — for fundamentally important reasons. The second because it led to a large, rich literature, and to a deep understanding of the nature of optimal play.[13]

Mixing *subjectively* means that rather than randomizing with "objective" devices like roulette wheels, where different players ascribe the same probability to any given outcome, the players use "subjective" devices, like horse races, where different players may ascribe different probabilities to the same outcome. Thus, a *subjective equilibrium* is defined as a profile of subjectively mixed strategies in which it is optimal for each player to play his strategy, given that the others play theirs. For this to make sense, it is not enough that different players ascribe different probabilities to the same event; they must also in some sense be mutually aware of those probabilities. Section 4.9j (pp. 108–110) grapples with this issue, inconclusively; inter alia, it says, "Though the players may know each other's probability, they may not know on precisely what observations it is based. Thus each player would not know if his own information is or is not more reliable than the other player's, and might be inclined to stick with his own information. Of course some

[13]See Chapter 7; also Aumann and Dreze (2008), "Rational Expectations in Games," *AER* 98, 72–86.

revision of probabilities would certainly be called for even in such a case; but whether such a revision must always ultimately lead to equal probabilities ... is not clear."

Two years later, the issue was settled with the publication of Chapter 6, which sets forth the idea of common knowledge and establishes the agreement theorem. These imply that subjective equilibrium can occur only when players have different priors. As a result, the idea of subjective equilibrium had no substantial echo, led nowhere, but the reason for its leading nowhere became a cornerstone of formal interactive epistemology.

The other generalization considered in Chapter 4 is *correlation*, in which one foregoes the requirement that the randomizations of different players be independent; thus, different players may get different information about the pure strategies played by all, and play accordingly. Formally, a *correlated equilibrium* is a random n-tuple of pure strategies, with distribution known to all players, in which each player is informed only of his own pure strategy, and it is optimal for him to play it, given that the others play theirs.

Underlying all three equilibrium notions — Nash, subjective, and correlated — is the notion of a player's *rationality*: that he does the best he can for himself — plays a pure strategy that is best for him,[14] given his beliefs. Chapter 7 demonstrates that if it is common knowledge among the players that all players are rational, then the random n-tuple of pure strategies that ensues is a correlated equilibrium. This holds only under the common prior assumption — that the players' probability assessments are based on their information only.

Subjective Probabilities

Probability assessments are central to Decision Theory in general, not only to Game Theory. In principle, there is no difficulty in defining probabilities for events such as a coin toss coming out heads, as coin tosses are repeatable. But what, exactly, do we mean when we say

[14]Yields the highest expected payoff.

that we attribute probability 1/3 to a given horse winning a given horse race?

In his 1954 book *The Foundations of Statistics*, Leonard Savage developed a definition of subjective probability based on axioms for preferences. Chapter 2, coauthored with Frank Anscombe, suggests an alternative approach, which proceeds from the familiar to the unfamiliar. Like Savage, we develop a definition of subjective probability based on conceptually compelling axioms for preferences, but unlike Savage, we distinguish explicitly between events based on repeatable processes such as coin tosses, for which probabilities are objectively defined, and those based on non-repeatable processes such as horse races. The upshot is that if p is your subjective probability of Nesara winning the fifth at Sunland Park, then you would as soon opt for a hundred dollars contingent on Nesara winning as for a hundred dollars with objective probability p. A priori, p might change if we substitute a thousand dollars for a hundred; our axioms imply that it does not.

Continuum of Players

According to the "law of supply and demand," prices in an unregulated economy adjust so that the total supply of each commodity equals the total demand for that commodity. However, this is subject to a caveat: The economy must have many agents, each of whom is "small" — has no significant effect on its workings.

In the late fifties and early sixties of the twentieth century, Martin Shubik, Herbert Scarf, and Gerard Debreu published a series of papers[15] that aim to make the above precise in game-theoretic terms; specifically, to relate the core[16] of the game defined by the economy with its *competitive* set — the set of those allocations arising from the law of supply and demand. The idea was to show that as the number

[15]See the references to Chapter 3, p. 70.

[16]See p. xiv above, the paragraph preceding the statement of the Strong Folk Theorem.

of agents grows indefinitely, while individual agents stay small,[17] the core approaches the competitive set. The difficulty was that both the core of an economy and its competitive set are subsets of spaces whose dimension depends on the number of agents, so it is not clear what is meant by "approaches."

During 1960–1961, when on sabbatical leave from the Hebrew University at Oskar Morgenstern's Econometric Research Program in Princeton, a Rand Corporation memorandum by John Milnor and Lloyd Shapley caught my attention. Entitled "Oceanic Games," it dealt with the Shapley values of voting games with a small number of relatively "large" players, who swim in an "ocean" — or continuum — of tiny, individually insignificant players. An example is a corporation with a few large stockholders and an "ocean" of small ones. In October of 1961, toward the end of my sabbatical, a conference on "Recent Advances in Game Theory" was held at Princeton. At that conference, Herb Scarf presented a theorem relating the core to the competitive set of economies with many small agents. Scarf's model had a denumerable infinity of traders, divided into a finite number of "types"; the model had some rough spots, for example, in that the total amount of goods is infinite. When hearing Scarf, I remembered Milnor and Shapley's oceanic games, and said to myself, "Surely, the continuum just has to be the right way to do this." The outcome is Chapter 5.[18]

Chapter 8 does for the Shapley value what Chapter 5 did for the core — shows that in exchange economies with a non-atomic continuum of agents, the nontransferable utility version of the Shapley value coincides with the competitive set.

---

[17] I.e., the maximum endowment of an agent stays small compared with the sum of all the endowments.

[18] Most of this paragraph is quoted from p. 157 of Volume II of my *Collected Papers*, ©2000 Massachusetts Institute of Technology, Cambridge, MA: The MIT Press.

Fair Division: Bankruptcy in the Talmud

While on sabbatical at Stanford during 1980–1981, I ran across a preprint by Barry O'Neill entitled "A Problem of Rights Arbitration from the Talmud." It aroused my interest, and I sent a copy to my son Shlomo (subsequently killed in action in one of Israel's wars), who was at that time studying at a Talmudical Academy in Jerusalem. In his response, Shlomo directed me to a fascinating Talmudic passage (Ketubot 93) related to that studied by O'Neill, but far more subtle. The passage proposes a division of the assets of a bankrupt debtor who owes a total of 600 zuz to three creditors: 100 to one, 200 to another, and 300 to a third. Three separate cases are considered: when the assets total 100, 200, or 300 zuz. On their face, the proposed divisions seem wildly inconsistent; indeed, with assets of 200, the logic of the division is hard to fathom even when considered on its own. For close to two millennia, Talmudic scholars did not succeed in coming up with a convincing explanation of this passage.[19]

When I returned to Jerusalem, Mike Maschler and I decided to see whether we couldn't bring the modern theory of games to bear on the problem. We tried many different game-theoretic approaches, but nothing worked. One day, one of us said, "Let's try the nucleolus; it can't hurt." The calculation (by hand) took about twenty minutes, and the result was startling: The nucleolus yielded precisely the nine strange numbers appearing in the Talmud.

Of course, this was only the beginning. The nucleolus is one of the more sophisticated game-theoretic solution concepts. Astute as they were, it is highly unlikely that the sages of the Talmud had anticipated Schmeidler's 1969 discovery of the nucleolus; they simply did not have the necessary technical equipment. What did seem possible was that there is some underlying principle common to both

[19]The early medieval authority Rabbi Hai Gaon (tenth century) did hint at the explanation offered below, and it is quite likely that he was aware of it. But he did not elaborate, and in subsequent years, this line of attack was abandoned. See p. 224 below.

the nucleolus and to the sages' way of thinking, which determines the division of assets in these bankruptcy problems. Mike and I set ourselves the task of finding such a principle.

A good place to look for "principles" is an axiomatization. In 1975, A. I. Sobolev published an axiomatization of the nucleolus using the principle of "consistency." This was the key to the puzzle, though it took many more months to work out. In the end, we were able to explain the Talmudic division on the basis of reasoning which, while relatively sophisticated, was well within the reach of the sages of the Talmud, and fits nicely with other Talmudic principles, and with the Talmudic passage itself.

For thousands of years, the commentators on the Talmud — who were much more astute than we — had been trying to make sense of this passage. Why do we think that we succeeded where they had failed? The reason is that modern game theory gave us a tool that they didn't have. Though the passage can be understood without game theory, that understanding would not have been *reached* without game theory.[20]

[20]Much of this section is quoted from pp. 60–61 of Volume II of my *Collected Papers*, ©2000 Massachusetts Institute of Technology, Cambridge, MA: The MIT Press.

About the Author

Robert Aumann was born in Frankfurt am Main, Germany, in 1930, to an orthodox Jewish family. Fleeing Nazi persecution, he emigrated to the United States with his family in 1938, settling in New York. In the process, his parents lost everything, but nevertheless gave their two children excellent Jewish and general educations. Aumann got a bachelor's degree from the City College of New York in 1950, and a Ph.D. in mathematics from MIT in 1955.

He joined the mathematics department at the Hebrew University of Jerusalem in 1956, and has been there ever since. In 1990, he was among the founders of the Center for the Study of Rationality at the Hebrew University, an interdisciplinary research center centered on Game Theory, with members from over a dozen different departments, including Business, Economics, Psychology, Computer Science, Law, Mathematics, Ecology, Philosophy, Statistics, and others.

Aumann is the author of over a hundred scientific papers and six books, and has held visiting positions at Princeton, Yale, Berkeley, Louvain, Stanford, Stony Brook, and NYU. He is a member of the American Academy of Arts and Sciences, the National Academy of Sciences (USA), the British Academy, and the Israel Academy of Sciences; holds honorary doctorates from the Universities of Chicago, Bonn, Louvain, City University of New York, Bar Ilan University, Ben Gurion University of the Negev, the University of Paris II Pantheon Assas, and Ariel University; and has received numerous prizes, including the Nobel Memorial Prize in Economic Sciences for 2005.

Aumann is married and had five children; the oldest was killed in action in the war between Israel and Lebanon in 1982. Also, he has twenty-one grandchildren, and as of this writing thirty-nine great-grandchildren, and one fifth-generation descendant. When not working, he likes to hike, cook, and study the Talmud.

Chapter 1

Acceptable Points in General Cooperative n-Person Games*

1.1. Introduction

In this paper, we introduce a new theory for n-person games. Like many previous theories for n-person games, it is based on the search for a reasonable "steady state" in the "supergame,"[1] i.e., the game each play of which consists of a long sequence of plays of the original game. We use two approaches: The usual approach, in which a single play is studied, and an attempt is made to find an n-tuple of strategies that intuitively speaking might be acceptable for a steady state in the supergame (Sections 1.4 and 1.5); and an approach based on a mathematical analysis of the supergame itself (Sections 1.6 and 1.7). The results we obtain by the two approaches turn out to be essentially equivalent (Sections 1.9 and 1.10).

We will consider here only cooperative games, i.e., games in which coalitions are permitted; side payments will be forbidden.[2] However, the ideas introduced here also apply to non-cooperative

*Reprinted from *Contributions to the Theory of Games IV* (1959), Annals of Mathematics Studies 40, edited by R. D. Luce and A. W. Tucker, 287–324, with permission from Princeton University Press.

[1]The name is due to Luce and Raiffa [6].

[2]Of course it is known (see [1]) that the cooperative game with side payments is a special case of the cooperative game without side payments.

games. Subsequent papers will consider non-cooperative as well as cooperative games.

We will not explicitly consider games involving chance; however, everything we say here carries over with no essential change to games involving chance.

Section 1.2 describes the notation we will use throughout. The reader may omit this section at first and use it only for reference. Section 1.8 is devoted to the proof of lemmas needed in the subsequent sections; it makes strong use of the approachability-excludability theory of Blackwell [4]. Sections 1.11, 1.12, and 1.13 are devoted to some miscellaneous remarks and counterexamples.

1.2. Notation

Throughout the paper, we will be concerned with an n-person game G. N will denote the set of all players; the individual players will be denoted by the positive integers $1, \ldots, n$. If x^i is an object defined for each $i \in N$, and if $B \subset N$, then x^B will denote the set of x^i where $i \in B$; it will be called a B-vector. When $B = N$, we will often omit the superscript and the prefix; that is, we will write x instead of x^N, and will refer to x simply as a vector rather than as an N-vector. If X is a set of vectors, we denote by X^B the set of all x^B where $x \in X$. When we speak of a subset B of N, i.e., of a $B \subset N$, we will mean a non-null subset, unless we specifically include the null subset. All B-vector equalities or inequalities will be taken to hold term by term.

We will be concerned only with random variables taking a fixed finite number of values, and possibly with infinite sequences of such random variables. If x is a random variable ranging over a finite set Z, then the probability distribution of x may be considered as a function y from Z to the reals satisfying the relations:

$$y(z) \geq 0, \quad z \in Z \tag{2.1}$$

and

$$\sum_{z \in Z} y(z) = 1. \tag{2.2}$$

We can then write

$$y = \sum_{z \in Z}^{*} y(z) z \tag{2.3}$$

where the $*$ indicates that the sum is merely symbolic and is not meant to be an ordinary sum. $y(z)$ is simply the probability that a random variable distributed according to y will take the value z.

The set of all probability distributions on Z will be denoted by $C(Z)$. The set of all pure strategies p^i for player i will be denoted by P^i. P is then the space of pure strategy vectors p. $C(P)$ is the space of *correlated* strategy vectors. A correlated strategy vector is a probability distribution on the set of all pure strategy *vectors*, just as a mixed strategy is a probability distribution on individual strategies. In general, to make use of a correlated strategy vector, the players have to agree to consult the *same* random device in choosing the pure strategies they will play. Correlated strategy vectors are thus not usable in non-cooperative games. We will usually abbreviate $C(P)$ to C. The prefix *c*- in front of a word will stand for "correlated."

Let X^B be a space of B-vectors, and let $B = \bigcup_{j=1}^{k} B_j$ be a partition of B into subsets. If for $j = 1, \ldots, k$, we have

$$y^{B_j} \in C(X^{B_j}),$$

then $(y^{B_1}, \ldots, y^{B_k})$ denotes the joint probability distribution of $y^{B_1}, \ldots, y^{B_k}$.

If Z_1 and Z_2 are finite sets and $y \in C(Z_1 \times Z_2)$, then $y|Z_1$ (sometimes also written $y|C(Z_1)$) will be the distribution given by

$$y|Z_1 = \sum_{z_1 \in Z_1}^{*} z_1 \sum_{z_2 \in Z_2} y(z_1, z_2).$$

If $B_1 \subset B_2$ and $y^{B_2} \in C(X^{B_2})$, then we will simply write y^{B_1} instead of $y^{B_2}|X^{B_1}$. This applies to all symbols *except* γ. Thus γ^{B_1} is *not* the same as $\gamma^{B_2}|P^{B_1}$. (γ^B will be defined on $C^B(P)$.)

Let Z_1 and Z_2 be finite sets, and suppose we have a function f that takes Z_1 into Z_2. If nothing is said to the contrary, we will assume that the definition of f is extended linearly; that is, f is a function from $C(Z_1)$ to $C(Z_2)$ defined by

$$f(y) = \sum_{z_1 \in Z_1}^{*} f(z_1)y(z_1).$$

If for each $i \in N$, we have a function f^i defined on X^i, then we may define a *vector function* f on X by means of

$$f(x) = (f^1(x^1), \dots, f^n(x^n)) \tag{2.4}$$

for all $x \in X$. It is important to remember, though, that the inverse process does not in general work. A vector function f defined on X does not yield a unique definition of a function f^i defined on X^i.

In general, when we have defined f^i on X^i, we will use the function f on X in the above sense.

For $p \in P$, the payoff to the ith player will be denoted by $E^i(p)$. If $c \in C$, $H(c)$ will be the mean of the distribution $E(c)$. $E(c)$ is called the *payoff distribution* when c is played, $H(c)$ the *expected* payoff when c is played. If $p \in P$, we have $H(p) = E(p)$.

The null set will be denoted by $\varnothing$.

We will be concerned with partitions of N. Let $\mathscr{P}_N$ denote the set of all subsets of N. Formally, a partition of N is a vector d whose components are members of $\mathscr{P}_N$, and which satisfies the condition

$$j \in d^i \Longleftrightarrow d^i = d^j, \quad i, j \in N.$$

The set of all partitions of N will be denoted by D. The *constant* partition d_N is defined by

$$d_N^i = N, \quad i \in N. \tag{2.5}$$

The *identity* partition d_e is defined by

$$d_e^i = \{i\}, \quad i \in N.$$

Denote by R the set of all vectors r, each of whose components r^i is a member of $\mathscr{P}_N$ containing i. Clearly $D \subset R$, and $D^i = R^i$. For a given d, the number of distinct d^i will be called $n(d)$. We choose a set of representatives $i_1, \ldots, i_{n(d)}$, one out of each distinct d^i. If c^B is a c-strategy B-vector, define

$$e(c^B) = B.$$

(The e stands for exponent; B is called the *exponent* of c^B.) Denote by T the set of all vectors t, each of whose components is some correlated strategy B-vector (not necessarily the same B for each member of N), for which

$$i \in e(t^i), \quad i \in N.$$

Let

$$d^i(t) = \{j | t^j = t^i\}, \quad i \in N, \quad t \in T. \tag{2.6}$$

Then $d(t)$ is a partition of N, and we have

$$d^i(t) \subset e(t^i), \quad i \in N.$$

Define

$$c(t) = ((t^{i_1})^{d^{i_1}(t)}, \ldots, (t^{i_{n(d(t))}})^{d^{i_{n(d(t))}}(t)}) \tag{2.7}$$

where $i_1, \ldots, i_{n(d(t))}$ are a set of representatives for $d(t)$.

In discussing games in extensive form, we will call the vertices of the game tree that are not moves by the name "terminal" rather than "play." This is necessary to avoid confusion, as we will be dealing with repeated plays of the same game, in the more common sense of the word.

As a final remark, we mention that everything we have stated probabilistically can be restated, where necessary, in the language of measure theory, and rigorously justified in that language. All sets we use can be proved to be measurable.

1.3. General Cooperative Games

We use the word "general" in the titles of the paper and of this section in order to emphasize that, unlike the cooperative games of von Neumann–Morgenstern, the games to be considered here exclude side payments. On the other hand, full and free cooperation is permitted between the players, in the sense that they may communicate with each other freely prior to each play and form any coalitions they may consider convenient. The agreements under which the coalitions are chosen are to be considered enforceable. (However, see Section 1.13 for further discussion of this point.) We will usually omit the word "general" and refer simply to a cooperative game. This is in accordance with the usage introduced by Nash [1].

In applications, the prohibition on side payments may be the result either of a physical barrier to side payments (in economic applications this might take the form of certain provisions in an anti-trust law) or of the lack of a common unit of measurement for the payoff. As has been pointed out by Raiffa [8], if there is a common unit of measurement for the payoff but there exists some physical barrier to the making of side payments, then the theory need only be invariant under linear transformations of the payoff function that do not destroy the common unit of measurement; i.e., under transformations of the form

$$H^{!} = aH + b \tag{3.1}$$

where both a and b are scalar multiples of the vector $(1, \ldots, 1)$ (a by a positive factor, b by an arbitrary factor). If, however, there is no common unit of measurement for the payoff, then the theory should be invariant under independent linear transformations of the payoff function; i.e., under transformations of the form (3.1), where a is an arbitrary positive vector and b is an arbitrary vector. In fact, the theory presented in this paper is invariant under the wider class of transformations, so that the results may be used in either kind of application.

Mathematically, the cooperative game is given by a vector payoff function that is assumed to be linear on the space C of correlated strategy vectors. The cooperative game differs from the non-cooperative game only in that the use of correlated strategy vectors is permitted in the former, but not in the latter.

This points up a reason for studying cooperative games even if our *a priori* interest lies in the field of non-cooperative games only.

Consider a long sequence of plays (we call this a *superplay*) of the game

	1	2
1	(4, 0)	(0, 0)
2	(0, 0)	(0, 4)

One of the most "natural" ways for the players to proceed in such a superplay would be to alternate between playing (1, 1) and (2, 2). This would tend to happen even in the non-cooperative case, by a sort of "silent gentlemen's agreement." However, the resulting payoff, (2, 2), is obtainable only by a correlated strategy vector, not by mixed strategies. Thus if we wish to arrive at results that are valid for the supergame by considering a single play, we must permit ourselves the use of correlated strategy vectors. This amounts to considering cooperative games.

1.4. Acceptable Points

Let c be a correlated strategy vector in G. Obviously, c will be a candidate for a possible steady state in the supergame unless some subset B of the players can, by concerted action, increase their payoff. Here it is not sufficient that the players of B be able to increase their payoffs by changing their strategies while the players of $N–B$ maintain their strategies as they are. For the players of $N–B$ will not in general maintain their strategies fixed in the face of a change on the part of B, but will also change *their* strategies in order to meet the new conditions; thus B's glory would be short-lived indeed. In order to rule out c as a candidate for a possible steady state, we have to be sure that the players of $N–B$ cannot prevent the players of B from improving their payoff by concerted action. Of course, we must require that *all* the members of B obtain more at the new point than they do at c, for they must all have an incentive to cooperate in the venture.

Definition 4.1. Let $c_0 \in C$. c_0 is said to be c-acceptable if there is no $B \subset N$ such that for each $c^{N-B} \in C^{N-B}$, there is a $c^B \in C^B$ for which

$$H^B(c^B, c^{N-B}) > H^B(c_0).$$

We write "c-acceptable" instead of "acceptable" because in subsequent papers of this series, we will be defining a slightly different kind of acceptability, and we will use prefixes to distinguish them. When in heuristic discussion in the sequel we use the word "acceptable" without a prefix, then it is to be taken to mean "acceptable" with an arbitrary prefix (but one that is fixed throughout the discussion). For the purposes of this paper alone, "acceptable" can be taken to mean the same thing as "c-acceptable."

The set of all c-acceptable c-strategy vectors is denoted by A_c. Note that the notion of acceptability is a "global" one. In other words, whether or not a point is acceptable depends only on the payoff at that point, not on the point itself. This is also easy to accept from the intuitive standpoint; as long as he is getting the same payoff, it makes little difference to a player which strategy he is playing.

Definition 4.2. A payoff vector h is said to be c-acceptable, if for some $c \in A_c$, we have

$$H(c) = h.$$

It will be convenient in the sequel to have available the following trivial restatement of 4.2.

Lemma 4.3. *A payoff vector h is c-acceptable if and only if for each $B \subset N$, there is a $c^{N-B} \in C^{N-B}$, such that for all $c^B \in C^B$, there is an $i \in B$ for which*

$$H^i(c^B, c^{N-B}) \geq h^i.$$

1.5. Two-Person, Zero-Sum Games

A sine qua non of theories of n-person games is that for two-person, zero-sum games, they reduce to the von Neumann theory. Because of the global nature of our theory, we can only expect that every acceptable payoff vector yield to each player the value of the game to that player. This is indeed the case.

Theorem 1. *In a two-person, zero-sum game G, a payoff vector h is c-acceptable if and only if*

$$h^1 = v, \tag{5.1}$$

where v is the value of G to Player 1.

Proof. Let h be c-acceptable. From Lemma 4.3 applied to $B = \{1\}$, we deduce that there is a $c_0^2 \in C^2$, such that for all $c^1 \in C^1$, we have

$$H^1(c^1, c_0^2) \leq h^1.$$

Hence

$$\max_{c^1 \in C^1} H^1(c^1, c_0^2) \leq h^1.$$

Hence

$$\min_{c^2 \in C^2} \max_{c^1 \in C^1} H^1(c^1, c^2) \leq h^1. \tag{5.2}$$

C^1 and C^2 are the mixed strategy spaces for Players 1 and 2 respectively. Hence the left side of (5.2) is v, and we deduce

$$v \leq h^1. \tag{5.3}$$

The value of G to Player 2 is $-v$. Proceeding as in the proof of (5.3), we obtain

$$-v \leq h^2. \tag{5.4}$$

But

$$h^2 = -h^1. \tag{5.5}$$

Combining (5.4) and (5.5), we obtain

$$v \geq h^1. \tag{5.6}$$

Combining (5.3) and (5.6), we obtain (5.1).

Conversely, assume (5.1). Then (5.3) and (5.2) follow at once. But for $B = \{1\}$, (5.2) is precisely the condition for c-acceptability given by Lemma 4.3. Similarly we establish the condition for $B = \{2\}$. To establish it for $B = \{1, 2\}$, note that for *any* (c^1, c^2), we have

$$H^1(c^1, c^2) + H^2(c^1, c^2) = 0 = h^1 + h^2,$$

whence we must have either

$$H^1(c^1, c^2) \leq h^1$$

or

$$H^2(c^1, c^2) \leq h^2.$$

Thus h is c-acceptable.

1.6. Strategies and Payoffs in the Supergame

In the notion of supergame that will be used in this paper, each superplay consists of an *infinite* number of plays of the original game G. On the face of it, this would seem to be unrealistic, but actually it is more realistic than the notion in which each superplay consists of a fixed finite (large) number of plays of G. In the latter notion, the fact that the players *know* when they have arrived at the last play becomes the decisive factor in the analysis, overshadowing all other considerations.[3] This is unnatural, because in the usual case, the players can always anticipate that there will still take place in the future an indefinite number of plays. This is especially true in political and economic applications, and even holds true for the neighborhood poker "club." Of course when looked at in the large, nobody really expects an infinite number of plays to take place; on the other hand, after each play we do expect that there will be more. A. W. Tucker has pointed out that this condition is mathematically equivalent to an infinite sequence of plays, so that is what our notion of supergame will consist of.

Roughly speaking, we will be interested in "behavior" strategies in the supergame; that is, we will be interested in strategy vectors

[3]See, for example, the excellent analysis of the supergame of the game known as the "Prisoner's Dilemma" that appears in 5.5 of [6].

that determine the c-strategy vector to be played on each play as a function of the information possessed by the players about the outcomes of the previous plays. Before we pass to the formal definition of a supergame c-strategy vector, let us examine in detail the procedure followed during the play of a cooperative supergame. The first question to be answered is: What information about the outcome of each previous play does each player i actually possess? The most he can know is exactly what pure strategy vector was actually played; the least he can be sure of knowing, in every game, is the strategy he himself played and the payoff he himself received. In between is a whole range of possibilities, depending on the rules of G. The situation can be concisely described by means of a function u^i that takes P onto some identification space of P. We define u^i so that for $p_1, p_2 \in P$, $u^i(p_1) = u^i(p_2)$ if and only if player i cannot tell after a play of G, whether the vector p_1 or the vector p_2 has been played. The minimum knowledge that each player i has, in accordance with our remarks above, insures us that if

$$u^i(p_1) = u^i(p_2),$$

then

$$p_1^i = p_2^i$$

and

$$H^i(p_1) = H^i(p_2).$$

Every game G has associated with it an information vector u satisfying these conditions. In particular, these conditions assure us of the existence of functions ϕ^i and ψ^i which map $u^i(P)$ onto P^i and $H^i(P)$ respectively. Some relations connecting the vector functions H, ϕ, ψ, and u are the following:

$$\phi \circ u = \text{identify}$$

$$\psi \circ u = H$$

$$\psi = H \circ \phi.$$

The information vector just defined is not included in the classical formulation of the extensive form of a game, as given, say, in [3]. That is because the state of information at the *end* of a game has heretofore been of little interest, as the payoff in no way depends on it. In a number of consecutive plays of a game, however, it is seen to be of vital importance. Incidentally, inclusion of this concept in the extensive structure of a game necessitates a revision in the notion of equivalence as formulated in [5]. We hope to treat the subject in more detail in a subsequent paper. For the present, we content ourselves with the remark that the information vector may be included in the extensive description of the game by associating with each player i, an information partition on the set of all terminals. In a game of perfect information, this partition would be the identity partition for each i.

All that we have said in this section up to now is quite general, applying to the non-cooperative as well as to the cooperative case. In the latter case, the information function does not exhaust the information that a player has to work with. Prior to each play, consultation takes place among the various players to determine the constitution of the coalitions and the correlated strategy B-vector that each coalition B will use. Formally, each player designates the coalition B to which he wishes to belong, and the correlated strategy B-vector he wishes B to play. Those players who can agree among each other form coalitions and play the agreed upon c-strategy B-vector.[4]

Naturally, although each player knows nothing about the coalitions in which the other players participate, he does remember the coalition B of which he himself was a member on previous plays.

[4]If $B_1 \subset B_2$ and every member of B_1 proposed (B_2, c^{B_2}) while the members of $B_2 - B_1$ proposed something else, then the members of B_1 form a coalition and play c^{B_1}.

The formal definition is as follows:

Definition 6.1. A supergame c-strategy f^i for player i is an infinite sequence of functions

$$f_0^i, f_1^i, \ldots, f_k^i, \ldots$$

where f_k^i is a function from

$$((U_1^i \times R_1^i) \times \cdots \times (U_k^i \times R_k^i))$$

to T^i; here U_j^i is a copy of $u^i(P)$ and R_j^i is a copy of R^i, for $j = 1, \ldots, k$.

The vector f is called a supergame c-strategy vector.

In the sequel, we will denote $U_j^i \times R_j^i$ by J_j^i. J_j^i represents the information about the jth play available to player i.

To define the payoff to a supergame c-strategy vector, we must first define the payoff to a given super*play*, as a function of its individual constituent plays. For this purpose, we use the arithmetic mean of the payoffs (in the limiting case, this amounts to the first Cesàro sum); Blackwell [4] also used the arithmetic mean under similar circumstances. Heuristically, one can argue both for and against this method; in my mind the heuristic advantages outweigh the disadvantages. Other methods have been proposed (see [6]) for obtaining the superplay payoff from the payoffs to its constituent plays; the most prominent seems to be the discounting of future payoffs back to the present, using a fixed rate of interest. This has the disadvantage of putting an unnaturally heavy stress on the beginning of the supergame. Nevertheless, it would be interesting to see an analysis based on this payoff notion, especially for certain kinds of economic applications.

A more perplexing problem in defining the payoff to a supergame c-strategy vector f is that of somehow combining the various possible superplays that might occur if f is played. The first impulse is to use some notion of *expected* payoff. We shall indeed define such a notion, but it is not an appropriate criterion for the players in choosing their supergame c-strategy. The chief reason for our being interested in expected payoffs for *individual* plays is that in a long sequence of plays, the law of large numbers turns an expected payoff into an *actual* payoff. A payoff for the whole superplay that is only *expected*, but not eventually approached, will not satisfy most players. Thus if a player has joined a coalition on the strength of a high expectation and is being disappointed in this expectation, he will soon express his disappointment by resigning from the coalition, regardless of his originally high expectation. For a clear example of the pitfalls of expected payoffs, see Section 1.12. What is required is some kind of criterion based on a law of large numbers. We give a restricted definition of this kind in this section, but reasoning of this type in general will dominate the remainder of the paper.

Let us first return to the question of expected payoff. For each $t \in T$, define

$$s(t) = (u(c(t)), d(t)). \tag{6.2}$$

Let f be a supergame c-strategy vector. For each $k \geq 1$, we may define a member $z_k(f) = z_k$ of $C(J_1 \times \cdots \times J_k)$ recursively as follows:

$$z_1 = s(f_0) \tag{6.3}$$

$$z_k = \sideset{}{^*}\sum_{y \in J_1 \times \cdots \times J_{k-1}} z_{k-1}(y)(y, s(f_{k-1}(y))), \quad k > 1. \tag{6.4}$$

z_k is the probability distribution of possible outcomes of plays up to the kth. Define a member $x_k(f) = x_k$ of $C(J_k)$ as follows:

$$x_k = s(f_{k-1}(z_{k-1})), \quad k \geq 1. \tag{6.5}$$

x_k is the probability distribution of outcomes of the kth play. Finally, define $E_k = E_k(f)$ and $H_k = H_k(f)$ by

$$E_k = E(c(f_{k-1}(z_{k-1}))) \tag{6.6}$$

and

$$H_k = H(c(f_{k-1}(z_{k-1}))). \tag{6.7}$$

E_k represents the probability distribution of possible payoff vectors on the kth play; H_k is the expected payoff for the kth play.

If

$$H(f) = \lim_{k \to \infty} \frac{1}{k} \sum_{r=1}^{k} H_r(f) \tag{6.8}$$

exists, then f is said to be *summable in the mean.*

Let $v = (v_1, \ldots, v_k, \ldots)$ be a sequence of random variables in which v_k is distributed according to the distribution $x_k(f)$. These random variables are not in general independent. Instead of saying that the random variables v_k are distributed according to $x_k(f)$, we will sometimes say that the random variable v is distributed according to f. The probability of a statement concerning a random variable v distributed according to f will usually be denoted Prob_f.

Definition 6.9. A supergame c-strategy vector f is said to be *summable* if it is summable in the mean and if a sequence of random variables distributed according to $E_k(f)$ obeys the strong law of large numbers.[5]

If v is an infinite sequence $(v_1, \ldots, v_k, \ldots)$ where $v_k \in J_k$ for all k, define

$$H_k(v) = \psi(v_k | U_k)(= H(\phi(v_k | U_k))) \tag{6.10}$$

and

$$S_k(v) = \frac{1}{k} \sum_{j=1}^{k} H_j(v). \tag{6.11}$$

Note that if v is a random variable distributed according to f, then $H_k(v)$ is a random variable distributed according to $E_k(f)$.

[5]See [2], p. 207. The law is there stated for one-dimensional random variables; the extension to n dimensions is straightforward.

Lemma 6.12. *A necessary and sufficient condition that a supergame c-strategy vector f be summable is that there exist a vector $H(f)$ such that for each vector $\varepsilon > 0$, we have*

$$\lim_{k\to\infty} \mathrm{Prob}_f(|S_r(v) - H(f)| \geq \varepsilon \text{ for some } r \geq k) = 0.$$

Furthermore, this $H(f)$ is then identical with the $H(f)$ defined by (6.8).

The proof, which is straightforward, will be omitted.

1.7. Strong Equilibrium Points

When we come to investigate what supergame c-strategy vectors f are liable actually to occur, we are faced with a problem somewhat different from the one we discussed in Section 1.4. There one could always look forward to future plays in order to punish deviations by the other players. In the supergame, however, there is only one superplay. Thus the players will have to make sure the first time that deviations won't pay. This amounts to saying that no coalition B can increase the payoff of all its members by substituting a different supergame c-strategy B-vector g^B while $N - B$ *maintains* the supergame c-strategy $(N - B)$-vector f^{N-B}. It is clear that such an f would be very strongly stable, since no *coalition* could have any incentive to move away from it, especially if we demand that B cannot even increase its payoff with positive probability, no matter how small. This is a much stronger version of what corresponds in the supergame to an ordinary equilibrium point.

The formal definition is as follows:

Definition. Let f be a summable supergame c-strategy vector. f is a *strong equilibrium c-point* if there is no $B \subset N$ for which there is a supergame c-strategy vector g satisfying

$$g^{N-B} = f^{N-B} \tag{7.1}$$

and

$$\lim_{k\to\infty} \mathrm{Prob}_g(S_r^B(v) \geq H^B(f) + \varepsilon^B \text{ for some } r \geq k) > 0 \tag{7.2}$$

for some B-vector $\varepsilon^B > 0$.

(The limit exists because the sequence involved is monotone decreasing.)

The set of all strong equilibrium c-points is denoted by S_c.

It is necessary to use the form $S_r^B(v) \geq H^B(f) + \varepsilon^B$ rather than $S_r^B(v) > H^B(f)$ because substitution of the latter statement would make condition 7.2 true even for $f = g$. We wish to include only those cases in which substitution of g for f will yield B an advantage that at least does not tend to the vanishing point, doubtful and rare as it may be.

Condition 7.2 is very weak, which means that the concept of strong equilibrium c-point is a very strong one (i.e., comparatively few supergame c-strategy vectors are strong equilibrium c-points). On the other end of the spectrum, we could replace 7.2 by

$$\lim_{k\to\infty} \mathrm{Prob}_g(S_r^B(v) \geq H^B(f) + \varepsilon^B \text{ for all } r \geq k) = 1 \quad \text{for some } B\text{-vector } \varepsilon^B > 0. \tag{7.3}$$

Denote by $\tilde{S}_c$ the set of supergame c-strategy vectors obtained when 7.3 is substituted for 7.2 in the definition of strong equilibrium c-point.

It is easy to see that 7.3 $\Longrightarrow$ 7.2, whence it follows that

$$S_c \subset \tilde{S}_c$$

and

$$H(S_c) \subset H(\tilde{S}_c). \tag{7.4}$$

We will demonstrate the opposite inequality as well. This shows that the notion of strong equilibrium point does not really depend on the strength of the condition 7.2.

1.8. Application of Approachability and Excludability Theory

For two-person, zero-sum games with vector payoffs, Blackwell [4] has defined the notion of "approachability" or "excludability" of a set in the payoff space with a supergame strategy employed by one or the other of the players. Blackwell does not use the information vector u which we introduced in Section 1.6, but apart from that, our Definition 6.1 applied to two-person, zero-sum games reduces to his definition of supergame strategy (for games not involving chance). Most of his definitions and results can be used here without any changes being necessary due to our use of the information vector.

Let us consider the two-person, zero-sum game with vector payoffs that is obtained from our game G by considering the coalition B to be the first player and the coalition $N - B$ the second player, the payoff to the first player being given by H^B.

Lemma 8.1. *Let $\Lambda(f, \varepsilon^B)$ denote the closed convex set consisting of all B-vectors λ^B for which*

$$\lambda^B \geq H^B(f) + \varepsilon^B.$$

Then if there is an $\varepsilon^B > 0$ such that $\Lambda(f, \varepsilon^B)$ is approachable with g^B, then 7.3 *holds.*

Proof. This lemma follows at once from the definition of approachability; see Section 1 of [4]. Our games do not involve chance, so the elements of the game matrix should be considered as B-vectors rather than probability distributions on B-vectors. We have abbreviated "approachable in the game matrix" to "approachable"; this will be done throughout the sequel.

Lemma 8.2. *If there is an $\varepsilon^B > 0$ such that, for each $c^{N-B} \in C^{N-B}$, there is a $c^B \in C^B$ for which*

$$H^B(c^B, c^{N-B}) \geq H^B(f) + \varepsilon^B,$$

then there is a g^B for which 7.3 holds.

Proof. The condition given in the lemma is equivalent to the statement that for each $c^{N-B} \in C^{N-B}, H^B(C^B, c^{N-B})$ intersects $\Lambda(f, \varepsilon^B)$. In accordance with Theorem 3 of [4], this in turn is equivalent to the statement that there is a g^B with which $\Lambda(f, \varepsilon^B)$ is approachable. Applying 8.1, we obtain the conclusion.

The information situation in [4] is slightly different from ours. In our application, the requirements of [4] may be stated as follows: Both B and $N-B$ know the payoffs to B on each previous play; nothing else is known. In our situation, B knows its own previous payoffs, but $N-B$ need not know what the payoffs to B were; on the other hand, both B and $N-B$ may have information given by the information function, that they do not have in [4]. An analysis of the proof of Theorem 3 of [4] shows that these differences do not affect the truth of the theorem, so that we may still apply it here.

To see this intuitively, note that the only difference that could interfere with the approachability of $\Lambda(f, \varepsilon^B)$ with g^B would be the additional information that $N-B$ has in our case that it does not have in [4]. Examination of the proof of Theorem 3 of [4] indicates that g_k^B depends only on the previous payoffs, and can be defined for *any* set of previous payoffs without affecting the truth of the theorem; thus no amount of additional information can possibly help $N-B$ in preventing B from approaching $\Lambda(f, \varepsilon^B)$ with g^B.

Lemma 8.3. *If there is a $c^{N-B} \in C^{N-B}$ (called γ^{N-B}) such that for each $c^B \in C^B$, there is an $i \in B$ for which*

$$H^i(c^B, \gamma^{N-B}) \leq H^i(f),$$

then for every B-vector $\varepsilon^B > 0$, and every supergame c-strategy vector β for which for all $(v_1, \ldots, v_k)$, we have

$$\beta_k^i(v_1, \ldots, v_k) = \gamma^{N-B}, \quad i \in N-B, \quad k \geq 0, \tag{8.4}$$

we have

$$\lim_{k\to\infty} \text{Prob}_\beta(S_r^B(v) \geq H^B(f) + \varepsilon^B \text{ for some } r \geq k) = 0.$$

Proof. The hypothesis of the lemma is equivalent to the assertion that for each $\varepsilon^B > 0$, $\Lambda(f, \varepsilon^B)$ fails to intersect $H(C^B, \gamma^{N-B})$. Applying Theorem 3 of [4], we conclude that for each $\varepsilon^B > 0, \Lambda(f, \varepsilon^B)$ is excludable with the supergame c-strategy $(N-B)$-vector given by (8.4). From the definition of excludability (Section 1 of [4]) it then follows that there is a number $\delta > 0$ such that

$$\lim_{k\to\infty} \text{Prob}_\beta \text{ (For all } r \geq k, \text{ there is an } i \in B \text{ such that}$$
$$S_r^i(v) < H^i(f) + \varepsilon^i - \delta) = 1,$$

whence

$$\lim_{k\to\infty} \text{Prob}_\beta(S_r^B(v) \geq H^B(f) + \varepsilon^B \text{ for some } r \geq k) = 0.$$

1.9. The Main Theorem: First Half

We will now show that our two approaches culminating in the definition of c-acceptability (4.1) and strong equilibrium (7.1, 7.2, and 7.3) actually yield results that are essentially the same. More precisely, for every c-acceptable strategy vector c, there is a strong equilibrium c-point (according to either 7.2 or 7.3) in supergame c-strategies that has the same payoff as c; and conversely. Thus c-acceptable points for a single play are seen to correspond exactly to strong equilibrium c-points in the supergame. The main theorem can be concisely stated by means of the equation

$$H(A_c) = H(S_c)(= H(\tilde{S}_c)).$$

This section is devoted to the proof of the relation

$$H(\tilde{S}_c) \subset H(A_c).$$

The proof rests heavily on Lemma 8.2. In addition to 8.2, the main fact needed is that if B cannot be prevented from obtaining more than h^B, then they cannot be prevented from obtaining a fixed amount more than h^B either. This is shown in Lemma 9.1.

Lemma 9.1. *Let h be a vector and B be a subset of N. If for each $c^{N-B} \in C^{N-B}$, there is a $c^B \in C^B$ for which*

$$H^B(c^B, c^{N-B}) > h^B,$$

then there is a positive B-vector ε^B such that for each $c^{N-B} \in C^{N-B}$, there is a $c^B \in C^B$ for which

$$H^B(c^B, c^{N-B}) \geq h^B + \varepsilon^B.$$

Proof. We first remark that if $G(x, y)$ is a continuous real-valued function of two variables x and y ranging over compact sets X and Y respectively, then the function defined by

$$F(y) = \max_{x \in X} G(x, y)$$

is continuous on Y.

Now define a function F on C^{N-B} by

$$F(c^{N-B}) = \max_{c^B \in C^B} \min_{i \in B}(H^i(c^B, c^{N-B}) - h^i).$$

By the above remark, F is continuous. Since C^{N-B} is compact, F takes on its minimum at a point c_0^{N-B} in C^{N-B}. By the hypothesis of the lemma,

$$F(c_0^{N-B}) > 0.$$

Setting

$$\varepsilon^i = F(c_0^{N-B})$$

for all $i \in B$, we obtain the conclusion of the lemma.

Theorem 2. $H(\tilde{S}_c) \subset H(A_c)$.

Proof. Let f be a summable supergame c-strategy vector; set

$$h = H(f).$$

Suppose

$$h \notin H(A_c), \tag{9.2}$$

i.e., suppose there is a B for which the hypothesis of Lemma 9.1 holds. Then the conclusion of Lemma 9.1, which is the same as the hypothesis of Lemma 8.2, holds. Hence we deduce the conclusion of

Lemma 8.2, i.e., there is a g^B for which 7.3 holds. Setting

$$g^{N-B} = f^{N-B},$$

we obtain a B and a g obeying 7.1 and 7.3, whence

$$f \notin \tilde{S}_c. \tag{9.3}$$

As we can go through the same argument for any f satisfying $h = H(f)$, we may deduce from (9.3) that

$$h \notin H(\tilde{S}_c). \tag{9.4}$$

We have shown that (9.2) $\Longrightarrow$ (9.4), whence we can deduce that

$$H(\tilde{S}_c) \subset H(A_c),$$

which is what we set out to prove.

1.10. The Main Theorem: Second Half

This section is devoted to the proof of the relation

$$H(A_c) \subset H(S_c);$$

together with Theorem 2, this yields the main theorem. The train of thought of the proof is somewhat as follows:

Suppose h to be a c-acceptable payoff. It is easy to set up a supergame c-strategy vector f whose payoff is h. But we must also incorporate in f a foolproof system for the punishment of any deviators; f must make sure that crime does not pay. The machinery for accomplishing this is at hand; by the definition of c-acceptability, for each B there is a c-strategy $(N-B)$-vector γ^{N-B} whose use prevents B from obtaining more than h^B. It remains only to *find* the culprits. To do this, we note that any players who at one time or another deviated from f were at the time of their deviation certainly not included in the same coalition as the "orthodox" players. Since each player knows who was in his coalition on all previous plays, the deviators are thus easily spotted. The set B_k of players who deviated on some play up to the $(k+1)^{\text{st}}$ is monotone increasing with k. Every time it grows, i.e., every time another player joins the ranks of the deviators, the remaining players revise their strategy to punish at least one of the *new* set of deviators. Since N is finite, there must be a set B that includes all the deviators and that is actually attained by B_k after a finite number of plays. The use of the strategy γ^{N-B} will then make certain that crime does not pay for at least one of the deviators. The proof is complicated by the fact that B_k and B usually depend on which pure strategies were chosen from among those considered by the correlated strategy vectors used on previous plays.

The strong equilibrium point f just described is one of "unrelenting ferocity"[6] against offenders. It exhibits a zeal for meting out justice that is entirely oblivious to the sometimes dire consequences to oneself or to the other faithful — i.e., those who have not deviated. There are other, more reasonable strong equilibrium points, which give deviating players a chance to return to the fold. These are important in connection with the general non-cooperative game; I will return to them in a subsequent paper on that topic. Here it is simpler and just as efficacious to use the strong equilibrium point of unrelenting ferocity.

Theorem 3. $H(A_c) \subset H(S_c)$.

Proof. Let $h \in H(A_c)$, and suppose $\gamma^N \in A_c$ is such that

$$H(\gamma^N) = h. \tag{10.1}$$

Then by 4.3, for each $B \subset N$ there is a $c^{N-B} \in C^{N-B}$, which we will call γ^{N-B}, such that for each $c^B \in C^B$, there is an $i \in B$ for which

$$H^i(c^B, \gamma^{N-B}) \le h^i. \tag{10.2}$$

[6]The term is due to Luce and Raiffa [6].

Now define a supergame c-strategy vector f as follows:

$$f_o^i = \gamma^N, \quad i \in N$$

$$f_k^i\ (v_1^i, \ldots, v_k^i) = \gamma^{v_k^i|R^i}, \ k > 0, \ i \in N, \ v_j^i \in J_j^i, \quad j \leq k. \qquad (10.3)$$

We must prove that the payoff to f is h and that f is a strong equilibrium c-point. It is convenient to break the proof into a number of lemmas. However, these lemmas will depend on previous assumptions and formulae, and are valid only within the context of this proof. They will be numbered like ordinary formulae used within the proof.

Lemma 10.4. *For $k > 0$ and $i \in N$, we have*

$$f_k^i(z_k) = \gamma^N.$$

Proof. The proof is by induction on k. For $k = 0$, (10.4) follows at once from (10.3). Assume (10.4) for a given k. Applying (2.6) and (2.7), we obtain

$$c(f_k(z_k)) = \gamma^N$$

and

$$d(f_k(z_k)) = d_N.$$

Hence, by (6.2),

$$s(f_k(z_k)) = (u(\gamma^N), d_N). \qquad (10.5)$$

Now by (6.4) and the linearity of f_k^i we have

$$f_{k+1}^i(z_{k+1}) = \sum_{y \in J_1 \times \cdots \times J_k}^{*} z_k(y) f_{k+1}^i(y, s(f_k(z_k)))$$

$$= \sum_{y \in J_1 \times \cdots \times J_k}^{*} z_k(y) f_{k+1}^i(y, (u(\gamma^N), d_N)) \quad \text{(by (10.5))}$$

$$= \sum_{y \in J_1 \times \cdots \times J_k}^{*} z_k(y) f_{k+1}^i(y^i, (u^i(\gamma^N), N))$$

(by (2.4) and (2.5))

$$= \sum_{y \in J_1 \times \cdots \times J_k}^{*} z_k(y) \gamma^N \qquad \text{(by (10.3))}$$

$$= \gamma^N \sum_{y \in J_1 \times \cdots \times J_k} z_k(y)$$

$$= \gamma^N. \qquad \text{(by (2.2))}$$

This completes the induction and the proof of the lemma.

Lemma 10.6. *For $k, j \geq 0$ and $k \neq j, c(f_k(z_k))$ is statistically independent of $c(f_j(z_j))$.*

Proof. By (10.3), the choice of strategies after each play depends only on the coalitions previously chosen. Thus the sequence of partitions of N into coalitions is strictly determined. For a given play, the choice of c-strategy B-vectors depends only on this sequence (which is, in fact, a sequence of constant partitions), and *not* on the strategies previously chosen. This completes the proof.

Lemma 10.7. *f is summable, and $H(f) = h$.*

Proof. Applying (2.7) to (10.4), we obtain

$$c(f_k(z_k)) = \gamma^N,$$

whence by (6.6)

$$E_k = E(c(f_k(z_k))) = E(\gamma^N). \qquad (10.8)$$

By Lemma 10.6, the strong law of large numbers[7] applies to each component of $c(f_k(z_k))$, hence to $c(f_k(z_k))$ itself, and hence also

[7]See [2], p. 207, *The Kolmogoroff Criterion*, or p. 208, the Theorem.

to E_k. By (10.8) and (6.7), we have

$$\begin{aligned} H_k &= H(\gamma^N) \\ &= h. \quad \text{(by (10.1))} \end{aligned}$$

Hence

$$\begin{aligned} H(f) &= \lim_{k\to\infty} \frac{1}{k}\sum_{r=1}^{k} H_r \quad \text{(by (6.8))} \\ &= \lim_{k\to\infty} \frac{1}{k} kh \\ &= \lim_{k\to\infty} h \\ &= h. \end{aligned}$$

By Definition 6.9, the proof is complete.

Lemma 10.9. *If $v_k|R$ is a partition and $k > 0$, then*

$$j \in e(f_k^i(v_1, \ldots, v_k)) \Longrightarrow f_k^j(v_1, \ldots, v_k) = f_k^i(v_1, \ldots, v_k).$$

Proof. From (10.3) and the left side of the conclusion, we deduce that $j \in v_k^i|R^i$. Hence since $v_k|R$ is a partition, it follows that

$$v_k^i|R^i = v_k^j|R^j,$$

and then the right side of the conclusion follows at once from (10.3).

In order to show that f is a strong equilibrium c-point, we will first assume that it is not.

Assumption 10.10. There is a $B \subset N$ (which we will call B^*) for which there is a supergame c-strategy vector g for which

$$g^{N-B^*} = f^{N-B^*} \tag{10.11}$$

and there is a B^*-vector $\varepsilon^{B^*} > 0$ for which

$$\lim_{k\to\infty} \text{Prob}_g(S_r^{B^*}(v) \geq H^{B^*}(f) + \varepsilon^{B^*} \text{ for some } r \geq k) > 0. \tag{10.12}$$

It will be our task to show that (10.10) leads to an absurdity.

Let $v = (v_1, \ldots, v_k, \ldots)$ be an arbitrary sequence for which $v_k \in J_k$, $k \geq 1$. Let $B(v)$ denote the set of i for which there is a $k_i \geq 0$ such that

$$g^i_{k_i}(v_1, \ldots, v_{k_i}) \neq f^i_{k_i}(v_1, \ldots, v_{k_i}). \tag{10.13}$$

It follows from (10.11) and (10.13) that

$$B(v) \subset B^*. \tag{10.14}$$

Definition 10.18. For each v and for $k \geq 0$, let $B_k(v)$ denote the set of i for which there is a k_i such that

$$0 \leq k_i \leq k$$

and such that (10.13) holds. Define

$$B_{-1}(v) = \varnothing. \tag{10.19}$$

For each v, there is clearly a $k(v)$ such that

$$B_k(v) = B(v), \quad k \geq k(v). \tag{10.20}$$

Furthermore, it is not difficult to see that (10.21): For each v, $B_k(v)$ is monotone increasing with k. We will in the sequel restrict use of the symbol $k(v)$ to the *first k(v)* satisfying (10.20).

Definition 10.22. We say that v occurs with *positive probability* if for each $k > 0$,

$$z_k(g)(v_1, \ldots, v_k) > 0;$$

that is, if for each $k > 0$, and under the assumption that the supergame c-strategy vector g is played, $(v_1, \ldots, v_k)$ occurs with positive probability.

Lemma 10.23. *If v occurs with positive probability, then for each $k > 0$, $v_k|R$ is a partition.*

Proof. By (6.4) and (10.22), $(v_1, \ldots, v_k)$ occurs with positive coefficient in

$$(v_1, \ldots, v_{k-1}, s(g_{k-1}(v_1, \ldots, v_{k-1}))).$$

Hence v_k occurs with positive coefficient in $s(g_{k-1}(v_1, \ldots, v_{k-1}))$. Applying (6.2), we conclude that $v_k|R$ occurs with positive coefficient in $d(g_{k-1}(v_1, \ldots, v_{k-1}))$. Since by definition, $d(g_{k-1}(v_1, \ldots, v_{k-1}))$ is a probability combination of a number of partitions of N, it follows that $v_k|R$ must be a partition, which completes the proof.

Lemma 10.24. *Let v occur with positive probability, and assume that* (10.10) *holds. For $k \geq 0$ and $i \in N - B_k(v)$, we have*

$$g_k^i(v_1, \ldots, v_k) = \gamma^{N - B_{k-1}(v)}.$$

Proof. We will write B instead of $B(v)$, and for each k, B_k instead of $B_k(v)$. The proof is by induction on k. First let $k = 0$. If $i \in N - B_0$, then it follows from Definition 10.18 that there is no k_i satisfying $k_i = 0$ and (10.13). In other words, for $i \in N - B_0$, we have

$$g_0^i = f_0^i.$$

Hence by (10.3), we have for $i \in N - B_0$,

$$\begin{aligned} g_0^i &= \gamma^N \\ &= \gamma^{N - B_{-1}} \text{ (by (10.19))}, \end{aligned}$$

which establishes (10.24) for $k = 0$.

Next, suppose we have established (10.24) for some $k \geq 0$ and all preceding k. For $i \in B_k - B_{k-1} = (N - B_{k-1}) \cap B_k$, we must have, by Definition 10.18, that (10.13) holds for $k_i = k$. Hence

$$g_k^i(v_1, \ldots, v_k) \neq f_k^i(v_1, \ldots, v_k), \quad i \in B_k - B_{k-1}. \tag{10.25}$$

Now if $k = 0$, it follows from (10.19), (10.25), and (10.3) that

$$g_0^i \neq \gamma^{N - B_{-1}}, \quad i \in B_0 - B_{-1}. \tag{10.26}$$

If $k > 0$, we have for $j \in B_k - B_{k-1}$ that

$$\begin{aligned} j &\in N - B_{k-1} \\ &= e(g_k^i(v_1, \ldots, v_k)), \quad i \in N - B_k \text{ (by induction hypothesis)} \\ &= e(f_k^i(v_1, \ldots, v_k)), \quad i \in N - B_k \text{ (by (10.18))}. \end{aligned}$$

Applying Lemma 10.9 and Lemma 10.23, we obtain

$$f_k^i(v_1, \ldots, v_k) = f_k^i(v_1, \ldots, v_k), \quad k > 0, \quad i \in N - B_k, \\ j \in B_k - B_{k-1}. \tag{10.27}$$

From (10.18) and the induction hypothesis, we obtain

$$f_k^i(v_1, \ldots, v_k) = g_k^i(v_1, \ldots, v_k) = \gamma^{N-B_{k-1}}, \quad i \in N - B_k, \ k > 0.$$

Combining this with (10.27), we obtain

$$f_k^j(v_1, \ldots, v_k) = \gamma^{N-B_{k-1}}, \quad j \in B_k - B_{k-1}, \ k > 0,$$

i.e.,

$$f_k^i(v_1, \ldots, v_k) = \gamma^{N-B_{k-1}}, \quad i \in B_k - B_{k-1}, \ k > 0.$$

From this and (10.25) we deduce, for $k > 0$, that

$$g_k^i(v_1, \ldots, v_k) \neq \gamma^{N-B_{k-1}}, \quad i \in B_k - B_{k-1}; \tag{10.28}$$

the same result for $k = 0$ is given by (10.26). Hence (10.28) holds for $k \geq 0$.

We certainly have

$$g_k^i(v_1, \ldots, v_k) \neq \gamma^{N-B_{k-1}}, \quad i \in B_{k-1},$$

for otherwise $i \notin e(g_k^i(v_1, \ldots, v_k))$, contradicting the membership of g_k^i in T^i. Combining this with the induction hypothesis and (10.28), we obtain

$$g_k^i(v_1,\ldots,v_k)\begin{cases}=\gamma^{N-B_{k-1}}, & i\in N-B_k\\ \neq\gamma^{N-B_{k-1}}, & i\in B_k.\end{cases}$$

It follows that

$$d^i(g_k(v_1,\ldots,v_k))=N-B_k,\quad i\in N-B_k,$$

and in particular, since $N-B_{k+1}\subset N-B_k$, we have

$$d^i(g_k(v_1,\ldots,v_k))=N-B_k,\quad i\in N-B_{k+1}. \tag{10.29}$$

Now for $i\in N-B_{k+1}$, we have by Definition 10.18 that

$$g_{k+1}^i(v_1,\ldots,v_{k+1})=f_{k+1}^i(v_1,\ldots,v_{k+1})=\gamma^{v_{k+1}^i|R^i}. \tag{10.30}$$

By reasoning similar to that used in the proof of (10.23), we obtain that $v_{k+1}^i|R^i$ occurs with positive coefficient in $d^i(g_k(v_1,\ldots,v_k))$. Combining this with (10.29) and (10.30), we obtain that for $i\in N-B_{k+1}$

$$g_{k+1}^i(v_1,\ldots,v_{k+1})=\gamma^{N-B_k}.$$

This completes the induction and establishes (10.24).

From (10.24) and (10.20) we deduce that if v occurs with positive probability, then

$$g_k^i(v_1,\ldots,v_k)=\gamma^{N-B(v)},\quad i\in N-B(v),\;\; k>k(v),$$

whence

$$c(g_k(v_1,\ldots,v_k))=(\gamma^{N-B(v)},\; c^{B(v)}(g_k(v_1,\ldots,v_k))),\quad k>k(v).$$

Setting

$$c^B(g_k(v_1,\ldots,v_k))=\gamma_k^B,\quad B\subset N,\;\; k>k(v),$$

we obtain

$$c(g_k(v_1,\ldots,v_k))=(\gamma_k^{B(v)},\; \gamma^{N-B(v)}),\quad k>k(v). \tag{10.31}$$

Lemma 10.32. *For each $m \geq 0$, there is a positive integer k, such that for all v, r, and $h_1, \ldots, h_m$ for which*

$$r \geq k \tag{10.33}$$

and

$$h_j \in H(P), \quad j = 1, \ldots, m, \tag{10.34}$$

and

$$S_r^{B^*}(v) \geq H^{B^*}(f) + \varepsilon^{B^*} \tag{10.35}$$

we have

$$r > m \tag{10.36}$$

and

$$\frac{1}{r}\left(\sum_{j=1}^{m} h_j^{B^*} + \sum_{j=m+1}^{r} H_j^{B^*}(v)\right) \geq H^{B^*}(f) + \frac{1}{3}\varepsilon^{B^*}. \tag{10.37}$$

Proof. First of all, we may choose

$$k > m,$$

so that (10.36) is satisfied. Next, choose

$$k \geq \max_{i \in B^*} \frac{3m}{\varepsilon^i} \max_{p \in P} H^i(p). \tag{10.38}$$

We then have

$$\left| S_r^{B^*}(v) - \frac{1}{r}\left(\sum_{j=1}^{m} h_j^{B^*} + \sum_{j=m+1}^{r} H_j^{B^*}(v)\right) \right|$$

$$= \left| \frac{1}{r}\left(\sum_{j=1}^{r} H_j^{B^*}(v)\right) - \frac{1}{r}\left(\sum_{j=1}^{m} h_j^{B^*} + \sum_{j=m+1}^{r} H_j^{B^*}(v)\right) \right| \quad \text{(by (6.11))}$$

$$= \frac{1}{r}\left|\sum_{j=1}^{m}(H_j^{B^*}(v) - h_j^{B^*})\right|$$

$$\leq \frac{1}{k}\sum_{j=1}^{m}\left|H_j^{B^*}(v) - h_j^{B^*}\right| \qquad \text{(by (10.33))}$$

$$\leq \frac{1}{k}\sum_{j=1}^{m} 2\max_{p\in P} H^{B^*}(p) \qquad \text{(by (10.34))}$$

$$= \frac{2m}{k}\max_{p\in P} H^{B^*}(p)$$

$$\leq \frac{2}{3}\varepsilon^{B^*} \qquad \text{(by (10.38))}$$

Combining this with (10.35), we obtain (10.37). This completes the proof.

Corollary 10.39. *For each $m \geq 0$, the following holds:*

For all sufficiently large k, we have that for all v, r, and $h_1, \ldots, h_m$ for which (10.33), (10.34), *and* (10.35) *hold,* (10.36) *and* (10.37) *follow.*

Proof. Because if (10.32) is true for a given k, it is certainly true for all larger k.

Lemma 10.40. *For each $m \geq 0$ and $B \subset N$, including the null set, we have*

$$\lim_{k\to\infty} \text{Prob}_g((S_r^{B^*}(v) \geq H^{B^*}(f) + \varepsilon^{B^*} \textit{ for some } r \geq k)$$

and $k(v) = m$ and $B(v) = B) = 0$.

Proof. First let $B = \varnothing$. Then for the v we are considering (inside Prob_g), we have

$$B(v) = \varnothing. \qquad (10.41)$$

Since $B_k(v)$ is monotone increasing with k and always contained in $B(v)$, it follows from (10.41) that for all k,

$$B_k(v) = \varnothing.$$

Applying (10.24), we obtain for $k \geq 0$ and $i \in N$,

$$g_k^i(v_1, \ldots, v_k) = \gamma^N. \tag{10.42}$$

Let θ be a supergame c-strategy vector given for all v by

$$\theta_k^i(v_1, \ldots, v_k) = \gamma^N, \quad k \geq 0, \quad i \in N. \tag{10.43}$$

From (10.42) and (10.43), we obtain

$$\begin{aligned}
&\lim_{k\to\infty} \text{Prob}_g((S_r^{B^*}(v) \geq H^{B^*}(f) + \varepsilon^{B^*} && \text{for some } r \geq k) \\
& && \text{and } k(v) = m \\
& && \text{and } B(v) = B) \\
&= \lim_{k\to\infty} \text{Prob}_\theta((S_r^{B^*}(v) \geq H^{B^*}(f) + \varepsilon^{B^*} && \text{for some } r \geq k) \\
& && \text{and } k(v) = m \\
& && \text{and } B(v) = B) \\
&\leq \lim_{k\to\infty} \text{Prob}_\theta(S_r^{B^*}(v) \geq H^{B^*}(f) + \varepsilon^{B^*} && \text{for some } r \geq k) \\
&\leq \lim_{k\to\infty} \text{Prob}_\theta(S_r^{i}(v) \geq H^{i}(f) + \varepsilon^{i} && \text{for some } r \geq k) \\
&= \lim_{k\to\infty} (1 - \text{Prob}_\theta(S_r^{i}(v) < H^{i}(f) + \varepsilon^{i} && \text{for all } r \geq k)) \\
&= 1 - \lim_{k\to\infty} \text{Prob}_\theta(S_r^{i}(v) < H^{i}(f) + \varepsilon^{i} && \text{for all } r \geq k) \\
&\leq 1 - \lim_{k\to\infty} \text{Prob}_\theta(|S_r^{i}(v) - H^{i}(f)| < \varepsilon^{i} && \text{for all } r \geq k),
\end{aligned} \tag{10.44}$$

where i is an arbitrary member of B^*.

Now when v is distributed according to θ, we deduce from (10.43) and (6.10) that the $H_j^i(v)$ are independent random variables whose mean is $H^i(\gamma^N)$. Applying (6.11), (10.1), (10.7), and the strong law of large numbers, we obtain that the second term of the right side of (10.44) is 1, whence we deduce that the left side vanishes. This completes the proof in the case $B = \varnothing$.

Next, suppose $B \neq \varnothing$. From (10.14) we obtain that $B(v) \subset B^*$. Thus if $B \not\subset B^*$, then $\mathrm{Prob}_g(B(v) = B) = 0$, and our lemma is already proved. Thus we may assume without loss of generality that

$$B \subset B^*. \tag{10.45}$$

For $j = 1, \ldots, m$, suppose the random variable w_j^B to be distributed according to an arbitrary but fixed member of $C^B(J_j^B)$, which we will call γ_j^B, while w_j^{N-B} is distributed according to γ^{N-B}. Set

$$h_j = H(w_j), \quad j = 1, \ldots, m. \tag{10.46}$$

Applying Corollary 10.39, we obtain

$$\begin{aligned}
\lim_{k\to\infty} \mathrm{Prob}_g((S_r^{B^*}(v) \geq H^{B^*}(f) + \varepsilon^{B^*} \quad & \text{for some } r \geq k) \\
& \text{and } k(v) = m \\
& \text{and } B(v) = B) \\
\leq \lim_{k\to\infty} \mathrm{Prob}\left(\left(\frac{1}{r}\left(\sum_{j=1}^{m} h_j^{B^*} + \sum_{j=m+1}^{r} H_j^{B^*}(v)\right) \geq H^{B^*}(f) + \frac{1}{3}\varepsilon^{B^*}\right.\right. & \\
& \text{for some } r \geq k) \\
& \text{and } k(v) = m \\
& \text{and } B(v) = B),
\end{aligned} \tag{10.47}$$

where $v_{m+1}, \ldots, v_r$ are distributed according to $x_{m+1}(g), \ldots, x_r(g)$, while h_j is given by (10.46). If for $j = k+1 \ldots, r$, we will let w_j be a random variable whose distribution is $(\gamma_j^B, \gamma^{N-B})$, where the w_j^B may be dependent on w_ℓ with $\ell < j$ but the w_j^{N-B} must be independent of all w_ℓ with $\ell \neq j$, then we may conclude from (10.31) and (6.10) that the right side of (10.47) is

$$\leq \lim_{k\to\infty} \text{Prob}\left(\frac{1}{r}\sum_{j=1}^{r} H^{B^*}(w_j) \geq H^{B^*}(f) + \frac{1}{3}\varepsilon^{B^*} \text{ for some } r \geq k\right)$$

$$\leq \lim_{k\to\infty} \text{Prob}\left(\frac{1}{r}\sum_{j=1}^{r} H^{B}(w_j) \geq H^{B}(f) + \frac{1}{3}\varepsilon^{B} \text{ for some } r \geq k\right) \tag{10.48}$$

by (10.45). Now there is certainly a supergame c-strategy vector β for which

$$c(\beta_{j-1}(z_{j-1}(\beta))) = (\gamma_j^B,\ \gamma^{N-B}), \quad j \geq 1,$$

whence the right side of (10.48)

$$= \lim_{k\to\infty} \text{Prob}_\beta\left(S_r^B(v) \geq H^B(f) + \frac{1}{3}\varepsilon^B \text{ for some } r \geq k\right). \tag{10.49}$$

Applying (10.2), (10.7) and Lemma 8.3, we obtain that the right side of (10.49) vanishes. Combining this with (10.48) and (10.47), we obtain the lemma.

Lemma 10.50.

$$\lim_{k\to\infty} \text{Prob}_g(S_r^{B^*}(v) \geq H^{B^*}(f)\ + \varepsilon^{B^*} \text{ for some } r \geq k) = 0.$$

Proof. Letting $\bigcup$ stand for mutually exclusive disjunction, we have

$$1 = \text{Prob}_g\left(\bigcup_{m=o}^{\infty} k(v) = m\right)$$

$$= \sum_{m=o}^{\infty} \text{Prob}_g(k(v) = m)$$

$$= \sum_{m=o}^{\infty} \text{Prob}_g \left(k(v) = m \text{ and } \bigcup_{B \subset N} (B(v) = B) \right)$$

$$= \sum_{m=o}^{\infty} \sum_{B \subset N} \text{Prob}_g(k(v) = m \text{ and } B(v) = B). \qquad (10.51)$$

Let $Q(v,k)$ stand for the expression

$$S_r^{B^*}(v) \geq H^{B^*}(f) + \varepsilon^{B^*} \text{ for some } r \geq k.$$

Then

$$\lim_{k \to \infty} \text{Prob}_g(Q(v,k))$$

$$= \lim_{k \to \infty} \text{Prob}_g(Q(v,k) \text{ and } \bigcup_{m=o}^{\infty} (k(v) = m) \text{ and } \bigcup_{B \subset N} (B(v) = B))$$

$$= \lim_{k \to \infty} \sum_{m=o}^{\infty} \sum_{B \subset N} \text{Prob}_g(Q(v,k) \text{ and } k(v) = m \text{ and } B(v) = B)$$

$$= \sum_{m=o}^{\infty} \sum_{B \subset N} \lim_{k \to \infty} \text{Prob}_g(Q(v,k) \text{ and } k(v) = m \text{ and } B(v) = B)$$

(since the series is dominated by the series on the right side of (10.51))

$$= \sum_{m=o}^{\infty} \sum_{B \subset N} 0 \quad \text{(by Lemma 10.40)}$$

$$= 0.$$

This completes the proof of Lemma 10.50.

Lemma 10.50 contradicts (10.12) and thus establishes the truth of Theorem 3.

Corollary 4. $H(A_c) = H(S_c) = H(\tilde{S}_c)$.

Proof. Combine (7.4), Theorem 2 and Theorem 3.

1.11. Existence of Acceptable Points

All two-person games, zero-sum or not, have acceptable points. The proof, which is not difficult, will be given in a subsequent paper of this series devoted exclusively to the application of this theory to two-person games. Unfortunately, though, when we go beyond two-person games, we find games in which there are no acceptable points. It is instructive to examine an example of such a game.

G is a three-person, zero-sum game. The three persons form a community and at regular intervals hold elections for mayor and vice-mayor. The mayor draws a salary of 2; the vice-mayor draws a salary of 1; the remaining player pays the salaries.

Let h be an arbitrary payoff vector. Since G is zero-sum, not all the players can have a positive payoff. Without loss of generality, let

$$h^1 \leq 0. \tag{11.1}$$

Now the minimum payoff to 1, even in pure strategies, is −3; hence since G is zero-sum, it follows that

$$h^2 + h^3 \leq 3. \tag{11.2}$$

From (11.2) it follows that at least one of 2 and 3 gets at most $1\frac{1}{2}$; so we may write without loss of generality

$$h^2 \leq 1\frac{1}{2}. \tag{11.3}$$

Now by agreeing to vote for 2 for mayor and for 1 for vice-mayor, 1 and 2 can both increase their payoffs; and there is nothing that 3 can do to prevent this.

Actually, the game G just described does not *deserve* an acceptable point. In a long sequence of plays of G, a steady state will never be reached; no sooner do we start settling down to one, than two of the players will see that it is relatively disadvantageous for them, and will certainly not be willing to agree to it for the remainder of the superplay. The game is *inherently unstable.* The same argument holds for all games not possessing acceptable points. Thus in seeking a steady state for the supergame, it would seem that we may restrict ourselves to games possessing acceptable points. These are called *stable* games.

1.12. The Expected Payoff

In order to see why the notion defined by (6.8) is essentially misleading, let us consider the three-person game G of perfect information which is given in extensive form in Figure 1.1.

The payoff vector at each terminal gives the payoff to players 1, 2, and 3 in that order.

Player 1 has three strategies, ℓ^1, c^1, and r^1 (left, center, and right). The other players have two strategies each; they will be denoted by ℓ^j and r^j, where $j = 0$, 2, and 3. Chance[8] is denoted by 0.

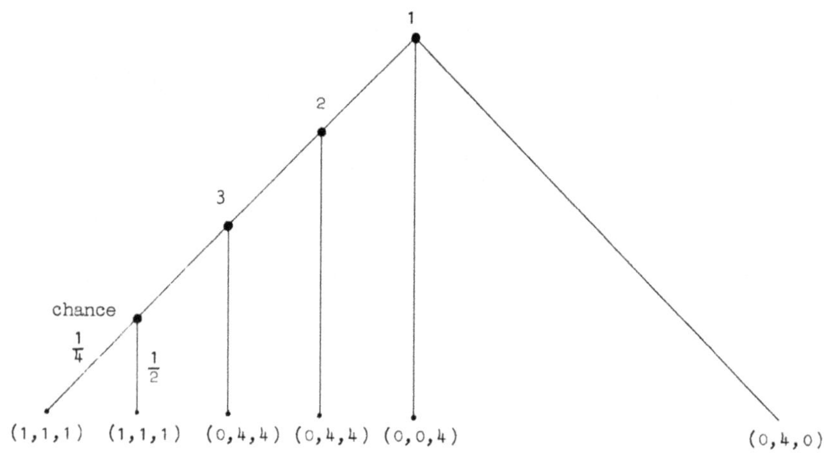

Figure 1.1: The payoff vector at each terminal gives the payoff to players 1, 2, and 3 in that order.

[8]It may be objected that we have excluded chance from consideration in this paper. Actually, chance is used in this example only to provide, as conveniently and cheaply as possible, a random device on which players 2 and 3 will be able to peg their choices. We could eliminate chance at the cost of complicating the example considerably. In order to exhibit the principle involved as clearly as possible, we have retained chance; but it is important to remember that it does not form an essential part of the example.

The point (ℓ^1, ℓ^2, ℓ^3), yielding a payoff of (1, 1, 1), is acceptable. No player or group of players can obtain a higher payoff. However, if Player 1 plays the supergame *c*-strategy given by (10.3), there is a supergame *c*-strategy (2,3)-vector that will yield an *expected* payoff of (0, 2, 2) rather than (1, 1, 1). This supergame *c*-strategy (2,3)-vector, which we will call the "plan," may be described as follows: On the first play, both players make the "orthodox" choice, i.e., they concur with Player 1 in choosing (ℓ^1, ℓ^2, ℓ^3). On all subsequent plays, we distinguish two cases:

1. If chance played ℓ^o on the first play, then Player 2 makes the "orthodox" choice, i.e., that given by (10.3), while Player 3 plays r^3.
2. If chance played r^o on the first play, then Player 3 makes the "orthodox" choice, while Player 2 plays r^2.

According to (10.3), Player 1 will respond to this treachery with a wounded roar and proceed to wreak dire vengeance. In case 1 he will finish off Player 3 by playing r^1; in case 2 he will play c^1, thus avenging himself on the treacherous Player 2. In each case, he is convinced that the orthodox player has been loyal all along, and therefore will not feel too unhappy about his getting 4 while Player 1 himself gets nothing. The expected payoff is (0, 2, 2), an improvement for both 2 and 3. For 2 and 3, this sounds like a good set-up, at least on paper. Who wouldn't want an expected payoff of 2 instead of 1? Let's see how it works out in practice.

Imagine that chance has played ℓ^o on the first play. For the second play, Player 1, all unsuspecting, suggests (ℓ^1, ℓ^2, ℓ^3), and Player 2 concurs, according to plan. Player 3 is now faced with the choice of going along with the plan and playing r^3, thus bringing down upon himself the eternal wrath of Player 1, or of concurring in the choice of (ℓ^1, ℓ^2, ℓ^3), thus possibly making Player 2 angry at him, but still assuring himself of a steady income of 1. It won't take him long to make his choice. He has nothing to gain and everything to lose by going along with the plan; he will get nothing but 0 for the rest of eternity for his pains, while Player 2 enjoys a steady payoff of 4. Player 3 will not now be consoled by the fact that if chance had played the other way, *he* would have been getting 4 while Player 2 would have been getting 0; in no sense does this constitute an incentive for him to follow the plan as things stand now. The point is that in a single play of a game, it may be worthwhile for each of the parties to plan to co-operate, even if it hurts, because in future plays, *he* may be the beneficiary. But in a superplay, there *are* no future superplays; *this is it*; by going along with the plan, Player 3 will be ruining himself not just for one play, but for always. It just isn't worthwhile, and no player would do it. Similar remarks apply to Player 2. The plan does yield a higher expected payoff, but will never be carried out.

It is possible to define strong equilibrium c-points using an appropriate formula involving expected payoffs in place of 7.2. If this is done, the first half of the main theorem, the analogue of Theorem 2, remains true.

However, the second half, the analogue of Theorem 3, fails. The game G that we have been discussing provides the counterexample. If we start out with $h = (1, 1, 1)$, then not only does the f given by (10.3) fail, as we have shown, but it can be shown that there is then no strong equilibrium c-point (in the sense of expected payoffs) whose payoff is h.

1.13. Enforceability of Agreements

There are two kinds of agreements involved in a cooperative supergame. One is the formal agreement that is contracted prior to a given play by the members of a coalition B, who agree to play a certain correlated strategy B-vector on that play. This agreement expires immediately after the play in question is completed; its validity does not extend to any subsequent plays. The other is the informal "gentlemen's agreement" in which the players undertake informally to make certain choices or join certain coalitions on each of a number of plays, or even for the entire superplay; they do not form a part of the formal mathematical structure introduced in Section 1.6. These are the agreements that are involved in the supergame strategy vectors that we have used in our proofs. When in heuristic discussions above we referred to "orthodoxy" or "deviationism," it was the second kind of agreement to which we were referring. To distinguish between the two kinds of agreements, we will call the first formal, the second informal.

As mentioned in Section 1.3, together with the formal agreement there comes a formal enforcement mechanism. Put completely precisely, if for the kth play the players choose the vector $t \in T$, then each $i \in N$ *must* become a member of $d^i(t)$, and the coalition $d^i(t)$ *must* then play $(t^i)^{d^i(t)}$. If the reader wishes, he may consider that once the choice of t has been made, the remainder of the play is no longer in the hands of the individual players, but in the hands of an umpire. Such an extreme interpretation is unnecessary for practical purposes; we refer to it only to make clear the intuitive meaning of formal enforcement.

On the other hand, the informal agreement can be enforced in no such formal way. Violation of such an agreement can be "prevented" only by the threat or implied threat of retaliation on subsequent plays. The exact nature of this "pseudo-enforceability" is of course the crux of what we have been investigating in the foregoing sections.

In the light of these remarks, it is an interesting fact that if in the supergame we abandon formal enforceability for the formal agreements, the main theorem remains true. The possibility of retaliation is sufficient to enforce even the formal agreements, without any necessity for formal enforcement apparatus. We have not proved this statement in the foregoing; the structure introduced in Section 1.6 was constructed with formal enforceability of formal agreements in mind. A structure permitting the violation of formal agreements would be somewhat more complicated.[9] However, upon a little reflection, the reader will be able to convince himself of the truth of the statement, at least intuitively.

---

[9]Roughly speaking, each play would be described by a choice of a $t \in T$ as in Section 1.6, followed by a chance choice of a pure strategy vector in accordance with $c(t)$, followed by some move permitting the players to renege on their commitments if they wish.

Of course, when considering just a single play rather than the supergame, as in the definition of acceptability, then there are only formal agreements and these are formally enforceable.

References

1. Nash, J. F., "Non-cooperative games," Annals of Mathematics 54 (1951), pp. 286–295.
2. Feller, W., An Introduction to Probability Theory and its Applications, John Wiley and Sons, 1950.
3. Kuhn, H. W., "Extensive games and the problems of information," Annals of Mathematics Study No. 28 (Princeton, 1953), pp. 193–216.
4. Blackwell, D., "An analog of the minimax theorem for vector payoffs," Pacific J. Math. 6 (1956), pp. 1–8.
5. Thompson, F. B., "Equivalence of games in extensive form," The RAND Corporation, Research Memorandum 759, 1952.
6. Luce, R. D., and Raiffa, H., Games and Decisions, John Wiley and Sons, 1957.
7. Von Neumann, J., and Morgenstern, O., Theory of Games and Economic Behavior, Princeton, 1944, 3rd ed., 1953.
8. Raiffa, H., "Arbitration schemes for generalized two-person games," Annals of Mathematics Study No. 28 (Princeton, 1953), pp. 361–387.

Chapter 2

A Definition of Subjective Probability*

2.1. Introduction

It is widely recognized that the word "probability" has two very different main senses. In its original meaning, which is still the popular meaning, the word is roughly synonymous with plausibility. It has reference to reasonableness of belief or expectation. If "logic" is interpreted in a broad sense, then this kind of probability belongs to logic. In its other meaning, which is that usually attributed to it by statisticians, the word has reference to a type of physical phenomena, known as random or chance phenomena. If "physics" is interpreted in a broad sense, then this kind of probability belongs to physics. Physical probabilities can be determined empirically by noting the proportion of successes in some trials. (The determination is inexact and unsure, like all other physical determinations.)

In order to distinguish these two main senses, physical probabilities will be referred to as "chances," whereas "probability" unqualified will refer to logical probability.

Within the two main categories of logical probability (probability proper) and physical probability (chances), especially in the former, various lesser differences of meaning can be distinguished. In this

*Co-authored with F. J. Anscombe. Reprinted from *Annals of Mathematical Statistics* (1963), 34(1), 199–205, with permission from Institute of Mathematical Statistics.

paper we are concerned with the personal or subjective concept of probability, as considered by Ramsey [13] and Savage [14]. Probabilities and utilities are defined in terms of a person's preferences, in so far as these preferences satisfy certain consistency assumptions. The definition is constructive; that is, the probabilities and utilities can be calculated from observed preferences.

Some persons, especially those with scientific training, are acquainted with the mathematical theory of chances and consider it to be an adequate theory for some kinds of physical phenomena — the uncertain outcomes of the spin of a roulette wheel, the toss of a coin, the roll of a die, a random-number generator. They believe that equipment can be found whose output conforms well with the theory of chances, with stochastic independence between successive observations, and *with stated values* for the chances of the simple outcomes. It suffices for this purpose that they believe that some one such piece of equipment exists, for example, a fair coin, since any system of chances can be realized by multiple use of such equipment. (The relation of the theory of chances with chance phenomena has been well illustrated by Kerrich [6]. See also Neyman [11].)

For such a person, his utilities can be defined in terms of chances, as shown by von Neumann and Morgenstern [10]. The purpose of this note is to define the person's probabilities in terms of chances, by an extension of the von Neumann–Morgenstern theory. The addition of only two plausible assumptions to those of utility theory permits a simple and natural definition of probabilities having the appropriate properties.

The idea of defining subjective probability in some such way has been "in the air" for a decade or so; in Section 2.5 we discuss some of the relevant literature. We believe that our presentation of subjective probability is the simplest so far given, for anyone who accepts the physical theory of chances.

2.2. Lotteries

All of the following considerations are based on the preferences of a single individual, whom we call "you."

Let a be a set of prizes. A *lottery* on a is a device for deciding which prize in a you will receive, on the basis of a single observation that records which one of a set of mutually exclusive and exhaustive uncertain events took place. It is possible that with each of these uncertain events there is associated a known chance; for example, this would be so if we were observing a single spin of a well-made roulette wheel. On the other hand, it is possible that chances cannot be associated with the uncertain events in question, or that the values of such chances are unknown; for example, this would be so if we were observing a horse race. To distinguish between the two kinds of lotteries, we call the first a "roulette lottery," and the second a "horse lottery." To fix ideas, we shall discuss only horse lotteries that are based on a single, particular horse race. Our object is to define the probabilities which you associate with each of the possible outcomes of this race.

In addition to the simple lotteries just described, we consider *compound* lotteries. These are constructed from simple lotteries by iteration; a lottery whose prizes are other lotteries is a compound lottery. A compound lottery may be compounded from roulette lotteries only, or from roulette lotteries and horse lotteries. Von Neumann–Morgenstern utilities of the basic prizes are constructed from preference comparisons between compound roulette lotteries. Subjective probabilities of the outcomes of the race will be constructed from preference comparisons between lotteries that are compounded from horse and roulette lotteries.

2.3. Assumptions

Utility theory for roulette lotteries based on a becomes mathematically easier if one assumes that among the prizes there is a most desired prize, say A_1, and a least desired prize, say A_0, such that A_1 is definitely preferred to $A_0 (A_1 \succ A_0)$ and for any prize A the following two "preference or indifference" relations hold: $A_1 \succsim A$, $A \succsim A_0$. This assumption has been made by Luce and Raiffa [9] — in fact, they go further and assume that a is finite, but that is unnecessary for their proofs — and we shall make it also.

Our notation for roulette lotteries will be the following: $(f_1B_1, \ldots, f_kB_k)$ denotes the roulette lottery that yields you B_1 with chance f_1, B_2 with chance f_2, and so on, where the f's sum to 1 and the B's stand for prizes from a or for lottery tickets of any sort under consideration. In respect of preferences, we shall regard $\boldsymbol{B}$ as equivalent to $(1B)$, for any B.

Let $\mathscr{R}$ be the set of all (simple or compound) roulette lotteries with prizes in a. We suppose that you have a preference ordering on $\mathscr{R}$ satisfying the axioms of utility theory (see for example [9]). Then it is possible to define a utility function u on $\mathscr{R}$ with the usual properties, including the following:

i. Associated with each prize A is a number $u(A)$, called the utility of A.
ii. If $B_1, \ldots, B_k$ are some prizes chosen from a, the simple roulette lottery $(f_1B_1, \ldots f_kB_k)$ has utility $f_1u(B_1) + \cdots + f_ku(B_k)$.
iii. Your preference ordering on $\mathscr{R}$ is isomorphic with the numerical ordering of the corresponding utilities. We may normalize the utility function so that $u(A_1) = 1, u(A_0) = 0$. For the proofs of these statements we refer the reader to Luce and Raiffa.

Turning now to the horse race, let us assume that it has exactly s mutually exclusive and exhaustive possible outcomes, denoted by $h_1, \ldots, h_s$. For any s roulette lotteries $R_1, \ldots, R_s$, chosen from $\mathscr{R}$, the symbol $[R_1, \ldots, R_s]$ will denote the (compound) horse lottery that yields you R_1 if the outcome of the race is h_1, R_2 if the outcome is h_2, and so on. The set of all such horse lotteries will be denoted by $\mathscr{H}$. Let $\mathscr{R}^*$ denote the set of all roulette lotteries *whose prizes are such horse lotteries*, i.e. members of $\mathscr{H}$ instead of members of a.

The essential device in our approach to subjective probability is to apply utility theory twice over, and then connect the two systems of preferences and utilities. We have just supposed that you have a preference ordering on $\mathscr{R}$, yielding the utility function u on $\mathscr{R}$. Let us now suppose that you also have a preference ordering on $\mathscr{R}^*$, satisfying the axioms of utility theory, with $\mathscr{H}$ taking the place of a. We shall distinguish these preferences by starred symbols $(\succsim^*, \succ^*)$, and denote the resulting utility function on $\mathscr{R}^*$ by u^*. We connect the two systems of preferences and utilities by the following two additional assumptions:

Assumption 1. (Montonicity in the prizes) *If* $R_i \succsim R'_i$, *then*

$$[R_1, \ldots, R_i, \ldots, R_s] \succsim^* [R_1, \ldots, R'_i, \ldots, R_s].$$

Assumption 2. (Reversal of order in compound lotteries)

$$(f_1[R_1^1, \ldots, R_s^1], \ldots, f_k[R_1^k, \ldots, R_s^k])$$
$$\sim^* [(f_1 R_1^1, \ldots, f_k R_1^k), \ldots, (f_1 R_s^1, \ldots, f_k R_s^k)].$$

Assumption 1 says that if two horse lotteries are identical except for the prizes associated with one outcome, then your preference between the lotteries is governed by your preference between the prizes associated with that outcome. It is very much akin to Luce and Raiffa's "substitutability" assumption. Assumption 2 says that if the prize you receive is to be determined by both a horse race and the spin of a roulette wheel, then it is immaterial whether the wheel is spun before or after the race. This is akin in spirit to Luce and Raiffa's "reduction of compound lotteries" assumption (sometimes called the "algebra of combining" assumption), but is even more plausible than the latter. Here the "joy in gambling" is not abstracted away, and the chances f_i are not combined in any way. Assumptions 1 and 2 reflect the intuitive idea that the outcome of the horse race is not affected by — nor does it affect — any spin of the roulette wheel.

2.4. Existence of Subjective Probabilities

From Assumption 1 it follows that $[A_1, \ldots, A_1]$ is the most desired, and $[A_0, \ldots, A_0]$ the least desired horse lottery. Let us normalize the utility function u^* on $\mathscr{R}^*$ so that $u^*[A_1, \ldots, A_1] = 1$ and $u^*[A_0, \ldots, A_0] = 0$.

Theorem. *There is a unique set of* s *non-negative numbers* $p_1, \ldots, p_s$ *summing to* 1, *such that for all* $[R_1, \ldots, R_s]$ *in* $\mathscr{H}$,

$$u^*[R_1, \ldots, R_s] = p_1 u(R_1) + \cdots + p_s u(R_s).$$

The number p_i is called the *subjective probability* of the outcome h_i of the race. From the theorem it follows that $u^*[R,\ldots,R] = u(R)$, so that we could identify $[R,\ldots,R]$ with R, and think of the preferences as constituting a single ordering defined on all lotteries simultaneously.

To prove the theorem, note first that by Assumption 1 any horse lottery is determined up to indifference by the utilities u of its entries; so that by abusing our symbolism slightly we can write our horse lotteries in the form $[r_1,\ldots,r_s]$, where $r_i = u(R_i)$. In particular, $u^*[1,\ldots,1] = u^*[A_1,\ldots,A_1] = 1$ and $u^*[0,\ldots,0] = u^*[A_0,\ldots,A_0] = 0$. Now define $p_i = u^*[0,\ldots,1,\ldots,0]$, where the 1 on the right side appears in spot i, and the other spots have 0.

Lemma. *If for some $k > 0$ and for $i = 1,\ldots,s$, we have $0 \leqq r_i \leqq 1$ and $0 \leqq kr_i \leqq 1$, then $u^*[kr_1,\ldots,kr_s] = ku^*[r_1,\ldots,r_s]$.*

To prove this, first assume $k \leqq 1$. Then by Assumption 2 and the expected utility property for utilities on $\mathscr{R}$, we have

$$\begin{aligned}[kr_1,\ldots,kr_s] &= [kr_1 + (1-k)0,\ldots,kr_s + (1-k)0] \\ &\sim^* (k[r_1,\ldots,r_s],\ (1-k)[0,\ldots,0]).\end{aligned}$$

Hence

$$\begin{aligned}u^*[kr_1,\ldots,kr_s] &= ku^*[r_1,\ldots,r_s] + (1-k)u^*[0,\ldots,0] \\ &= ku^*[r_1,\ldots,r_s].\end{aligned}$$

Next, if $k > 1$, it follows from the first half that

$$u^*[r_1,\ldots,r_s] = u^*[kr_1/k,\ldots,kr_s/k] = (1/k)u^*[kr_1,\ldots,kr_s];$$

and multiplying through by k, we complete the proof of the lemma.

Returning to the proof of the theorem, we set $c = r_1 + \cdots + r_s$. If $c = 0$, then all the $r_i = 0$, and the theorem is trivial. If $c > 0$, then the r_i/c are nonnegative and sum to 1, and hence from Assumption 2 we conclude that

$$[r_1/c,\ldots,r_s/c] \sim^* ((r_1/c)[1,0,\ldots,0],\ldots,(r_s/c)[0,\ldots,0,1]).$$

Hence by the lemma (with $k = 1/c$) and the expected utility property, we have

$$\begin{aligned} u^*[R_1, \ldots, R_s] &= u^*[r_1, \ldots, r_s] \\ &= u^*[cr_1/c, \ldots, cr_s/c] = cu^*[r_1/c, \ldots, r_s/c] \\ &= c((r_1/c)u^*[1, 0, \ldots, 0] + \cdots + (r_s/c)u^*[0, \ldots, 0, 1]) \\ &= r_1p_1 + \cdots + r_sp_s = p_1u(R_1) + \cdots + p_su(R_s), \end{aligned}$$

and the proof of the theorem is complete.

If our construction of subjective probabilities is applied to a set of exclusive and exhaustive outcomes h_i of some trial, such that each outcome has a known chance f_i, the "horse lotteries" degenerate into "roulette lotteries." (Formally this means that we are assuming $[R_1, \ldots, R_s] \sim (f_1R_1, \ldots, f_sR_s)$.) We now see at once from our definition of p_i that $p_i = f_i$ for all i. Thus in this case the subjective probability of any outcome is equal to the chance associated with that outcome. Since the two are equal, it does not matter much which word or symbol we use. The chance refers to the phenomenon, the probability refers to your attitude towards the phenomenon, and they are in perfect agreement.

Reverting now to proper horse lotteries for which there are no chances, we can interpret our construction of probabilities thus: p_i is equal to the chance such that you would as soon opt for a dollar contingent on the ith horse's winning as for a dollar with that chance. *A priori* this chance might change if we substituted two dollars for a dollar, but our theorem says that it does not.

To provide an adequate basis for the study of scientific inference, the above development of probability needs some extension. Horses must be translated into hypotheses and lotteries into decisions, and the concept of observation must be introduced. Though this involves no particular mathematical difficulty, it does raise some philosophical questions. We do not pursue the matter here.

The above theory of horse lotteries can be extended from a finite set of outcomes $\{h_i\}$ to a measurable space of outcomes. We refrain from going into details.

2.5. Comparison with the Literature

Ramsey [13], de Finetti [4], Koopman [7], Good [5], Savage [14], Davidson and Suppes [3], Suppes [15], Kraft, Pratt and Seidenberg [8] and others have given definitions of subjective probability. Most often this has been done without reference to the physical theory of chances. Some writers, no doubt, have felt that because physics bristles with philosophical difficulties it is undesirable to base a logical concept on a physical concept; and that once a logical concept of probability has been adequately defined there is no need to contemplate any separate physical concept of chance. But it is possible to hold an opposite view, that probability is even obscurer than chance and that progress should preferably be from the more familiar to the less familiar, rather than the other way round. At the least, this route deserves to be explored. Blackwell and Girshick [1], Chernoff [2], Raiffa and Schlaifer [12] and others have explored it and obtained results closely related to those given here. The novelty of our presentation, if any, lies in the double use of utility theory, permitting the very simple and plausible assumptions and the simple construction and proof.

A comparison of our terminology and approach with Savage's [14] may be helpful. Our "horse lottery" corresponds to his "act"; our "outcome of the race" to his "state of the world"; our "prize" to his "consequence." Of Savage's six postulates, which he numbers P1 through P6, we share with him explicitly P1 (ordering of the horse lotteries — it is among the assumptions of utility theory). We also share P5, which Suppes [15] terms a structure axiom rather than a rationality axiom; it is our assumption that $A_1 \succ A_0$. P2 and P3 (the "sure-thing" principle) are represented by our Assumption 1, which is close to P3. P4 can be translated to our context by saying that your guess about which horse will win is not affected by the size of the prize offered for each horse. Inherent in the intuitive background of P4 is the assumption that you explicitly guess which horse will win, and that you are able and willing to judge which of two horses has a higher probability of winning. Thus P4 implicitly assumes the existence of comparative probabilities; rather than constructing subjective probabilities from preferences, Savage has constructed numerical subjective probabilities from comparative subjective probabilities. There is nothing corresponding to P4 in our approach. P6 is a continuity postulate which plays a role in defining subjective probability similar to that of the continuity (or "Archimedean") assumption of utility theory in defining the utility function. Here again there is implicit use of the probability notion; but now we have not only comparative probabilities, but also the existence of probabilities that are "small." Although there is nothing directly corresponding to P6 in our approach, we may say, very roughly, that there is no need for such an assumption in our approach because we start out with chances, which have a continuous range.

It would seem that a postulate asserting the existence of comparative probabilities, like Savage's P4, occurs in many presentations of subjective probability. For example, the requirement L_3 of Blackwell and Girshick [1], which is rather sophisticated and includes Savage's P2 as a special case, includes also as a special case something similar in spirit to Savage's P4, though not identical with it. We feel that the preference relation is a step closer to experience than the comparative probability relation, and that it is desirable to base subjective probabilities on the preference relation only, without any *a priori* reference — implicit or explicit — to comparative probabilities.

In this respect our treatment resembles that of Raiffa and Schlaifer [12]. They, however, claim only that their treatment is "*very* informal." It appears to us that they make use of what we call Assumption 2 without mentioning it. In this paper, Assumption 2 is certainly not a tautology, as can easily be shown by counterexample. However, in a different context, with different basic assumptions, our Assumption 2 *could* be a tautology. It is indeed very obvious, intuitively.

To conclude, we remark that anyone who wishes to avoid a concept of physical chance distinct from probability may reinterpret our construction as a method of defining more difficult probabilities in terms of easier ones. Such a person may consider that probabilities can be assigned directly to the outcomes of spins of a roulette wheel, flips of a coin, and suchlike, from considerations of symmetry. The probabilities may be so widely agreed on as to be termed impersonal or objective probabilities. Then, with some assumptions concerning independence, our construction can be used to define subjective probabilities for other sorts of outcomes in terms of these objective probabilities.

In fact, apropos of the above remark and of the preceding comments on comparative probabilities, Professor Savage has pointed out in private correspondence that the incidence of P4 in the axiom system of [14] can be weakened, so that P4 is asserted only for a subset of acts, corresponding (roughly) to our roulette lotteries. He has then deduced the original P4. Thus an unrestricted postulate of comparative probabilities is not essential to the derivation of numerical probabilities from Savage's viewpoint.

Acknowledgments

Research partly supported by the Office of Naval Research, in the Department of Mathematics and in the Econometric Research Program, Princeton University. We are indebted to Professor Savage for much valuable discussion, and to the referee for the reference to Chernoff.

References

1. Blackwell, D. and Girshick, M. A. (1954). *Theory of Games and Statistical Decisions.* Wiley, New York. Section 4.3.
2. Chernoff, H. (1954). Rational selection of decision functions. *Econometrica* **22** 422–443. Section 7.
3. Davidson, D. and Suppes, P. (1956). A finitistic axiomatization of subjective probability and utility. *Econometrica* **24** 264–275.
4. De Finetti, B. (1937). La prévision: ses lois logiques, ses sources subjectives. *Ann. Inst. H. Poincaré* **7** 1–68.
5. Good, I. J. (1950). *Probability and the Weighing of Evidence.* Griffin, London.
6. Kerrich, J. E. (1946). *An Experimental Introduction to the Theory of Probability.* Munksgaard, Copenhagen.
7. Koopman, B. O. (1940). The axioms and algebra of intuitive probability. *Ann. of Math.* **41** 269–292.
8. Kraft, C. H., Pratt, J. W. and Seidenberg, A. (1959). Intuitive probability on finite sets. *Ann. Math. Statist.* **30** 408–419.

9. Luce, R. D. and Raiffa, H. (1957). *Games and Decisions.* Wiley, New York. Section 2.5.
10. Neumann, J. von and Morgenstern, O. (1947). *Theory of Games and Economic Behavior*, 2nd ed. Princeton University Press. Chapter 3 and Appendix.
11. Neyman, J. (1952). *Lectures and Conferences on Mathematical Statistics and Probability*, 2nd ed. U.S. Department of Agriculture Graduate School, Washington. Chapter I, Parts 1 and 2.
12. Raiffa, H. and Schlaifer, R. (1961). *Applied Statistical Decision Theory.* Harvard University Graduate School of Business Administration, Boston. Section 1.5.
13. Ramsey, F. P. (1931). Truth and probability. *The Foundations of Mathematics and Other Logical Essays.* Routledge and Kegan Paul, London. 156–198.
14. Savage, L. J. (1954). *The Foundations of Statistics.* Wiley, New York. Chapters 2 and 3.
15. Suppes, P. (1956). The role of subjective probability and utility in decision-making. *Proc. Third Berkeley Symp. Math. Statist. Prob.* **5** 61–73. University of California Press.

Note added in proof Professor T. S. Ferguson has pointed out that our application of utility theory to horse lotteries in Section 2.4 depends on the assumption of strict preference,

$$[A_1, \ldots, A_1] \succ^* [A_0, \ldots, A_0].$$

This would follow at once from our previous assumption that $A_1 \succ A_0$ if we made the identification of $[R, \ldots, R]$ with R.

Chapter 3

Markets with a Continuum of Traders*

It is suggested that the most natural mathematical model for a market with "perfect competition" is one in which there is a continuum of traders (like the continuum of points on a line). It is shown that the core of such a market coincides with the set of its "equilibrium allocations," i.e., allocations which constitute a competitive equilibrium when combined with an appropriate price structure.

3.1. Introduction

The notion of *perfect competition* is fundamental in the treatment of economic equilibrium. The essential idea of this notion is that the economy under consideration has a "very large" number of participants, and that the influence of each individual participant is "negligible." Of course, in real life no competition is perfect; but, in economics, as in the physical sciences,[1] the study of the ideal state has proved very fruitful, though in practice it is, at best, only approximately achieved.

Though writers on economic equilibrium have traditionally assumed perfect competition,[2] they have, paradoxically, adopted

*Reprinted from *Econometrica* (1964), 32(1/2), 39–50, with permission from Wiley.

[1]Think of a "freely falling body" (no air resistance), an "ideal gas" (the molecules do not collide), an "ideal fluid" (incompressible and nonviscous), and so on.

[2]A discussion of the literature will be found in Section 3.5.

a mathematical model that does not fit this assumption. Indeed, the influence of an individual participant on the economy cannot be mathematically negligible, as long as there are only finitely many participants. Thus *a mathematical model appropriate to the intuitive notion of perfect competition must contain infinitely many participants.* We submit that the most natural model for this purpose contains a *continuum* of participants, similar to the continuum of points on a line or the continuum of particles in a fluid. Very succinctly, the reason for this is that one can integrate over a continuum, and changing the integrand at a single point does not affect the value of the integral, that is, the actions of a single individual are negligible.

This paper is confined to "pure exchange economies," i.e., markets. Presumably, the results could be extended to economies with production, but we have not done this. We investigate and compare two known concepts of market equilibrium — the *competitive equilibrium* and the *core*. By means of the continuous model, it is possible to express the relation between these two concepts in a particularly simple and transparent manner. In fact, in the continuous model — but in no finite model — the two concepts are essentially equivalent.

The market model that we consider consists of a set of traders, each of whom starts out with an initial commodity bundle to be used for trading, and each of whom has a well-defined preference order on the set of all commodity bundles. A *trade* (or *allocation*) is a redistribution of the commodities in the initial bundles among the traders. This is a perfectly standard model in economic theory with the sole exception that the set of traders has heretofore been assumed finite.

A *competitive equilibrium* is a state of the market arrived at via "the law of supply and demand"; it consists of a price structure p (one price for each commodity) at which the total supply of each good exactly balances the total demand, and the allocation $\boldsymbol{x}$ that results from trading at these prices. More precisely, $\boldsymbol{x}$ is an allocation with the property that at the price structure p, no trader can, with the value of his initial bundle, buy a bundle that he prefers to his part of $\boldsymbol{x}$. If $(p, \boldsymbol{x})$ is a competitive equilibrium, then $\boldsymbol{x}$ is called an *equilibrium allocation.*

An allocation $\boldsymbol{x}$ is said to be in the *core* of the market if no coalition of traders can force an outcome that is better for them than $\boldsymbol{x}$. More precisely, $\boldsymbol{x}$ is in the core if there is no group of traders that, by its own efforts alone, without help from traders not in the group, can assure each of its members of a final commodity bundle preferred to that obtained under $\boldsymbol{x}$. What we mean by "its own efforts" is that the desired result can be obtained if the traders in the group exchange the commodities in their initial bundles among themselves only, as if the other traders were not present. The core is a generalization of Edgeworth's "contract curve" [**8**].

It is widely recognized that the notion of competitive equilibrium makes economic sense only if perfect competition is assumed. Otherwise, a change in an individual's offer to buy or sell can easily upset prevailing prices, so that the restriction to these prices is meaningless. The notion of core, on the other hand, does not depend on perfect competition; it is perfectly valid even for markets containing only two or three traders.

The definition of competitive equilibrium assumes that the traders allow market pressures to determine prices and that they then trade in accordance with these prices, whereas that of core ignores the price mechanism and involves only direct trading between the participants. Intuitively, one feels that money and prices are no more than a device to simplify trading, and therefore the two concepts should lead to the same allocations. It is to be expected that this will not happen in finite markets, where the notion of competitive equilibrium is not really applicable; and, indeed, though every equilibrium allocation is always in the core, the core of a finite market usually contains points that are not equilibrium allocations. But, when the notion of perfect competition *is built into the model*, that is, in a continuous market, one may expect that the core equals the set of equilibrium allocations. That this is indeed the case is the main result of this paper.

It has long been conjectured that some such theorem holds; the basic idea dates back at least to Edgeworth [**8**]. The usual rough statement is that "the core approaches the set of equilibrium allocations as the number of traders tends to infinity." Unfortunately, it is extremely difficult to lend precise meaning to this kind of statement, to say nothing of proving it. Very recently, Debreu and Scarf did succeed in stating and proving a theorem of this kind in a brilliant and elegant fashion [**7**]; even so, their theorem holds only under comparatively restrictive conditions. Their work will be discussed and compared with ours in Section 3.5. There we will also discuss an earlier model of Scarf [**18**] and Debreu [**6**], containing a denumerable infinity of traders.

Continuous models are nothing new in economics or game theory, but it is usually parameters such as price or strategy that are allowed to vary continuously. Models with a continuum of *players* (traders in this instance) are a relative novelty, and the references can still be counted on the fingers of one hand. Milnor and Shapley [**15**] and Shapley alone [**19**] pioneered the idea in two papers dealing with power indices (Shapley values); [**19**] is set in a context of considerable economic interest. The only other references of which we know are Davis [**4**] and Peleg [**17**], who treat such games from the point of view of their von Neumann-Morgenstern solutions and their "bargaining sets," respectively.

The idea of a continuum of traders may seem outlandish to the reader. Actually, it is no stranger than a continuum of prices or of strategies or a continuum of "particles" in fluid mechanics. In all these cases, the continuum can be considered an approximation to the "true" situation in which there is a large but finite number of particles (or traders or strategies or possible prices). The purpose of adopting the continuous approximation is to make available the powerful and elegant methods of the branch of mathematics called "analysis," in a situation where treatment by finite methods would be much more difficult or even hopeless (think of trying to do fluid mechanics by solving n-body problems for large n).

There is perhaps a certain psychological difference between a fluid with a continuum of particles and a market with a continuum of traders. Though we are intellectually convinced that a fluid contains only finitely many particles, to the naked eye it still looks quite continuous. The economic structure of a shopping center, on the other hand, does not look continuous at all. But, for the economic policy maker in Washington, or for any professional macroeconomist, there is no such difference. He works with figures that are summarized for geographic regions, different industries, and so on; the individual consumer (or merchant) is as anonymous to him as the individual molecule is to the physicist.

Of course, to the extent that individual consumers or merchants are in fact *not* anonymous (think of General Motors), the continuous model is inappropriate, and our results do not apply to such a situation. But, in that case, perfect competition does not obtain either. In many real markets the competition is indeed far from perfect; such markets are probably best represented by a mixed model, in which some of the traders are points in a continuum, and others are individually significant (compare [**19**]). The purpose of this paper is to study the extreme case in which perfect competition does obtain, i.e., there are no individually significant traders.

It should be emphasized that our consideration of a continuum of traders is not merely a mathematical exercise; it is the expression of an economic idea. This is underscored by the fact that the chief result holds *only* for a continuum of traders — it is false for any finite number. It would presumably also be possible to consider a continuum of different commodities, but, in contrast to the continuum of traders, this would serve no useful purpose (in this context). A continuum of commodities is the appropriate mathematical idealization of a real situation in which there are "many" commodities, but our result holds for any number of commodities, many or few, so there is nothing to be gained by considering only the case of "many" commodities. The continuum of traders, on the other hand, captures an idea and enables the proof of a theorem that could *not* be proved in the finite set-up.

Our chief mathematical tools are Lebesgue measure and integration, but only their most elementary properties are needed. Riemann integration can *not* be substituted (see the beginning of the proof of Lemma 4.2). Much of the proof is adapted from the work of Debreu and Scarf [**6, 7**], but becomes simpler and more natural in this context. Indeed, over and above the specific result obtained here, what we would like to stress is the power and simplicity of the continuum-of-players method in describing mass phenomena in economics and game theory. The present work should be considered primarily as an illustration of this method as applied to an area where no other treatment seemed completely satisfactory, and we hope that it may stimulate more extensive development and use of models with a continuum of players.

The mathematical model and the main theorem are presented in the following section, and the assumptions are briefly discussed in Section 3.3. Section 3.4 is devoted to the proof of the main theorem. Section 3.5 reviews the literature and compares the present work with that of other authors.

3.2. The Mathematical Model and the Main Theorem

We will be working in a Euclidean space R^n; the dimensionality n of the space represents the number of different commodities being traded in the market.

Superscripts will be used exclusively to denote coordinates. Following standard practice, for x and y in R^n we write $x > y$ to mean $x^i > y^i$ for all i; $x \geqq y$ to mean $x^i \geqq y^i$ for all i; and $x \geq y$ to mean $x \geqq y$ but not $x = y$. The integral of a vector function is to be taken as the vector of integrals of the components. The scalar product $\Sigma_{i=1}^n x^i y^i$ of two members x and y of R^n is denoted $x \cdot y$. The symbol 0 denotes the origin in R^n as well as the real number zero; no confusion will result.

A *commodity bundle* x is a point in the nonnegative orthant Ω of R^n. The set of *traders* is the closed unit interval [0, 1]; it will be denoted T. An *assignment* (of commodity bundles to traders) is a function $\boldsymbol{x}$ from T to Ω, each coordinate of which is Lebesgue integrable over T. As all integrals are with respect to t, we shall omit writing t, as well as the symbol dt, under the integral sign; thus $\int_T \boldsymbol{x}$ means $\int_T \boldsymbol{x}(t)dt$.

There is a fixed *initial assignment* $\boldsymbol{i}$. Intuitively, $\boldsymbol{i}(t)$ is the bundle with which trader t starts out. We assume

$$\int_T \boldsymbol{i} > 0. \tag{2.1}$$

An *allocation* (or "final assignment" or "trade") is an assignment $\boldsymbol{x}$ for which

$$\int_T \boldsymbol{x} = \int_T \boldsymbol{i}.$$

For each trader t there is defined a relation $\succ_t$ on Ω, which is called the *preference relation* of t and satisfies the following conditions:

Desirability (of the commodities): $x \geqq y$ implies $x \succ_t y$. (2.2)

Continuity (in the commodities): For each $y \in \Omega$, the sets $\{x\colon x \succ_t y\}$ and $\{x\colon y \succ_t x\}$ are open (relative to Ω). (2.3)

Measurability: If $\boldsymbol{x}$ and $\boldsymbol{y}$ are assignments, then the set $\{t : \boldsymbol{x}(t) \succ_t \boldsymbol{y}(t)\}$ is Lebesgue measurable in T. Note specifically that $\succ_t$ is not assumed to be complete, nor even transitive. (2.4)

A *coalition* of traders is a Lebesgue measurable subset of T; if it is of measure 0, it is called *null*.[3] An allocation $\boldsymbol{y}$ *dominates* an allocation $\boldsymbol{x}$ *via* a coalition S if $\boldsymbol{y}(t) \succ_t \boldsymbol{x}(t)$ for each $t \in S$, and S is *effective* for $\boldsymbol{y}$, i.e.,

$$\int_S \boldsymbol{y} = \int_S \boldsymbol{i}.$$

The *core* is the set of all allocations that are not dominated via any nonnull coalition.

[3]Not to be confused with the *empty* coalition, which has no members at all. The term "almost every trader" will mean "every trader except possibly for a null set."

A *price vector* p is an n-tuple of nonnegative real numbers, not all of which vanish. A *competitive equilibrium* is a pair consisting of a price vector p and an allocation $\boldsymbol{x}$, such that for almost every trader t, $\boldsymbol{x}(t)$ is maximal with respect to $\succ_t$ in t's *budget set* $\{\boldsymbol{x}\colon p \cdot x \leqq p \cdot \boldsymbol{i}(t)\}$. An *equilibrium allocation* is an allocation $\boldsymbol{x}$ for which there exists a price vector p such that $(p, \boldsymbol{x})$ is a competitive equilibrium.

Theorem. *The core coincides with the set of equilibrium allocations.*

3.3. Discussion of the Model

The definitions of allocation, core, and competitive equilibrium are the same as the usual ones, except that where the usual definitions sum over a set of traders, we integrate. It is useful to think of the commodity bundle $\boldsymbol{x}(t)$ held by an individual trader t under an assignment $\boldsymbol{x}$ as an "infinitesimal." Only when a "significant" (i.e., nonnull) coalition pools its resources can a non-infinitesimal bundle result, namely $\int_S \boldsymbol{x}$. Thus an individual trader t — and, in general, any null coalition — is without influence in the market, and can and will always be ignored. Intuitively, this is what justifies the exclusion of null coalitions in the definitions of core and competitive equilibrium.

The intuitive meaning of "S is effective for $\boldsymbol{y}$" is that the coalition S can assure to each of its members t the bundle $\boldsymbol{y}(t)$, without any help from traders not in S.

This it can do by an appropriate redistribution of the initial bundles $\boldsymbol{i}(t)$ because of the condition $\int_S \boldsymbol{y} = \int_S \boldsymbol{i}$ that defines effectiveness.

Assumption (2.1) asserts that each of the commodities is actually present in the market. In other words, each commodity is initially held by some traders, though different commodities might be held by different traders, and it could well happen that no single trader initially holds a positive amount of all commodities. The finite analogue of this assumption has been used by McKenzie [**5**, p. 58, Assumption 5]. Though very weak, the assumption is not merely a normalization; it can be shown that the main theorem is false without it. On the other hand, for a given real-life market, it is always possible and, in fact, natural to build a model satisfying (2.1), namely, by counting as commodities in the model only those commodities actually present in the real-life market.

The desirability assumption (2.2) says that each trader always wants more of every commodity. One consequence is that satiation is never reached.[4] The continuity assumption (2.3) is the usual one; it asserts that if $x \succ_t y$, and x' and y' are sufficiently close to x and y, respectively, then also $x' \succ_t y'$. The measurability assumption (2.4) is of technical significance only and constitutes no real economic restriction. Nonmeasurable sets are extremely "pathological"; it is unlikely that they would occur in the context of an economic model.

We remark that our assumptions are far weaker than those usually assumed in market models.

The choice of the unit interval as a model for the set of traders is of no particular significance. A planar or spatial region would have done just as well. In technical terms, T can be any measure space *without atoms*.[5] The condition that T have no atoms is precisely what is needed to ensure that each individual trader have no influence.

3.4. Proof of the Main Theorem

First, we show that every equilibrium allocation is in the core. Let $(p, \boldsymbol{x})$ be a competitive equilibrium. Suppose, contrary to the theorem, that $\boldsymbol{x}$ is dominated via a coalition S by an allocation $\boldsymbol{y}$. Then, by the definition of competitive equilibrium, we have $p \cdot \boldsymbol{y}(t) > p \cdot \boldsymbol{i}(t)$ for almost all $t \in S$. Hence

$$p \cdot \int_S \boldsymbol{y} = \int_S p \cdot \boldsymbol{y} > \int_S p \cdot \boldsymbol{i} = p \cdot \int_S \boldsymbol{i},$$

[4]The assumption can be dispensed with for $y > \boldsymbol{i}(t)$; this would permit satiation for such y.

[5]An *atom* of a measure space is a nonnull set in the space which includes no nonnull subset of smaller measure. The best example is a "mass point," i.e., a single point that carries positive measure.

and this contradicts

$$\int_S \boldsymbol{y} = \int_S \boldsymbol{i}.$$

Conversely, we show that every allocation in the core is an equilibrium allocation. Let $\boldsymbol{x}$ be in the core. Define

$$\boldsymbol{F}(t) = \{x\colon x \succ_t \boldsymbol{x}(t)\},$$
$$\boldsymbol{G}(t) = \boldsymbol{F}(t) - \boldsymbol{i}(t) = \{x - \boldsymbol{i}(t)\colon x \in \boldsymbol{F}(t)\}.$$

For each set U of traders, let $\Delta(U)$ denote the convex hull of the union $\cup_{t\in U}\boldsymbol{G}(t)$. Define U to be *full*, if its complement is null.

Lemma 4.1. *There is a full set U of traders, such that 0 is not an interior point of $\Delta(U)$.*

Proof. For each x in R^n, let $\boldsymbol{G}^{-1}(x)$ be the set of all traders t for whom $\boldsymbol{G}(t)$ contains x. The notation $\boldsymbol{G}^{-1}$ is suggested by the fact that $t \in \boldsymbol{G}^{-1}(x)$ if and only if $x \in \boldsymbol{G}(t)$. From $\boldsymbol{G}^{-1}(x) = \{t\colon x + \boldsymbol{i}(t) \succ_t \boldsymbol{x}(t)\}$ and measurability (2.4), it follows that $\boldsymbol{G}^{-1}(x)$ is measurable for each x.

Let N be the set of all those rational points r in R^n (i.e., points with rational coordinates) for which $\boldsymbol{G}^{-1}(r)$ is null. Obviously N is denumerable. Define $U = T\backslash\cup_{r\in N}\,\boldsymbol{G}^{-1}(r)$, where $\backslash$ denotes set-theoretic subtraction. Then U is full.

Suppose 0 is in the interior of $\Delta(U)$. Then there is a point $x > 0$ such that $-x \in \Delta(U)$; by definition of $\Delta(U)$, $-x$ is a convex combination of finitely many points in $\cup_{t\in U}\boldsymbol{G}(t)$. That is, there are traders $t_1,\ldots,t_k \in U$ (not necessarily distinct), points $x_i \in \boldsymbol{G}(t_i)$, and positive numbers $\beta_1,\ldots,\beta_k$ summing to 1, such that $\Sigma_{i=1}^k\beta_i x_i = -x < 0$. By the continuity assumption (2.3), we may find rational points $r_i \in \boldsymbol{G}(t_i)$ sufficiently close to the x_i, and positive rational numbers γ_i sufficiently close to the β_i, so that we still have $\Sigma_{i=1}^k\gamma_i r_i < 0$. Let $-r = \Sigma_{i=1}^k\gamma_i r_i$, and pick an arbitrary trader t_0 in U. Since $r > 0$, we have $\alpha r + \boldsymbol{i}(t_0) > \boldsymbol{x}(t_0)$ for sufficiently large positive rational α. Hence by the desirability assumption (2.2), $\alpha r + \boldsymbol{i}(t_0) \succ_t \boldsymbol{x}(t_0)$, i.e., $\alpha r \in \boldsymbol{G}(t_0)$. Now set $r_0 = \alpha r, \alpha_0 = 1/(\alpha+1), \alpha_i = \alpha\gamma_i/(\alpha+1)$ for $i = 1,\ldots,k$. Then

$\alpha_i > 0$ for all i, and $\Sigma_{i=0}^k \alpha_i = 1$; furthermore

$$\sum_{i=0}^{k} \alpha_i r_i = \frac{\alpha}{\alpha+1} r + \frac{\alpha}{\alpha+1} \sum_{i=1}^{k} \gamma_i r_i = 0,$$

and $r_i \in \boldsymbol{G}(t_i)$ for all i. Then $t_i \in \boldsymbol{G}^{-1}(r_i)$, and, since $t_i \in U$, it follows that $r_i \notin N$. Therefore $\boldsymbol{G}^{-1}(r_i)$ is of positive measure for each i. Therefore, for a sufficiently small positive number δ, we can find disjoint subsets S_i of $\boldsymbol{G}^{-1}(r_i)$ such that $\mu(S_i) = \delta\alpha_i$, where μ is Lebesgue measure. Define a coalition S by $S = \cup_{i=0}^k S_i$, and an assignment $\boldsymbol{y}$ by

$$\boldsymbol{y}(t) = \begin{cases} r_i + \boldsymbol{i}(t) & \text{for } t \in S_i, \\ \boldsymbol{i}(t) & \text{for } t \notin S. \end{cases}$$

That $\boldsymbol{y}(t) \in \Omega$ is trivial for $t \notin S$, and for $t \in S_i$ it follows from $r_i \in \boldsymbol{G}(t)$ (which in turn follows from $S_i \subset \boldsymbol{G}^{-1}(r_i)$). Next,

$$\int_S \boldsymbol{y} = \sum_{i=0}^{n} \delta\alpha_i r_i + \int_S \boldsymbol{i} = 0 + \int_S \boldsymbol{i},$$

and therefore S is effective for $\boldsymbol{y}$; since $\boldsymbol{y}(t) = \boldsymbol{i}(t)$ for $t \notin S$, it follows that $\boldsymbol{y}$ is an allocation. Finally, from $S_i \subset \boldsymbol{G}^{-1}(r_i)$ it follows that $r_i + \boldsymbol{i}(t) \succ_t \boldsymbol{x}(t)$ for $t \in S_i$; in other words, $\boldsymbol{y}(t) \succ_t \boldsymbol{x}(t)$ for $t \in S$. Since S is of positive measure, we have shown that $\boldsymbol{x}$ is not in the core, contrary to assumption. This proves the lemma.

Let U be as in the lemma. To avoid annoying repetitions, let us agree that in the remainder of the proof, statements about traders will refer to $t \in U$. This is sufficient because U is full.

From the lemma and the supporting hyperplane theorem we obtain a hyperplane $p \cdot x = 0$ that supports $\Delta(U)$. Therefore it also supports each of the $\boldsymbol{G}(t)$; thus $p \cdot x \geqq 0$ for $x \in \boldsymbol{G}(t)$, or

$$p \cdot x \geqq p \cdot \boldsymbol{i}(t) \qquad \text{for } x \in \boldsymbol{F}(t). \tag{4.2}$$

By desirability (2.2), each $\boldsymbol{F}(t)$ contains a translate of the positive orthant; therefore $p \geqq 0$. We shall show that $(p, \boldsymbol{x})$ is a competitive equilibrium.

We first show that for almost all t, $\boldsymbol{x}(t)$ is in t's budget set, i.e.,

$$p \cdot \boldsymbol{x}(t) \leqq p \cdot \boldsymbol{i}(t). \tag{4.3}$$

Indeed, because of desirability (2.2), there are bundles arbitrarily close to $\boldsymbol{x}(t)$ that t prefers to $\boldsymbol{x}(t)$. Therefore $\boldsymbol{x}(t)$ is in the closure of $\boldsymbol{F}(t)$, and so, by (4.2), $p{\cdot}\boldsymbol{x}(t) \geqq p{\cdot}\boldsymbol{i}(t)$. If, for a nonnull t-set, we would have $p \cdot \boldsymbol{x}(t) > p \cdot \boldsymbol{i}(t)$, then it would follow that $\int_T p \cdot \boldsymbol{x} > \int_T p \cdot \boldsymbol{i}$, contrary to the assumption that $\boldsymbol{x}$ is in the core and hence is an allocation. This demonstrates (4.3) for almost all t; we may and will assume that all traders satisfy it.

To complete the proof, we must show that $\boldsymbol{x}(t)$ is maximal in t's budget set, i.e., that (4.2) can be sharpened to

$$p \cdot x > p \cdot \boldsymbol{i}(t) \qquad \text{for } x \in \boldsymbol{F}(t). \tag{4.4}$$

To this end, we first establish $p > 0$. Suppose not; let $p^1 = 0$, say. Since $p \neq 0$, some coordinate of p does not vanish; let $p^2 > 0$, say. By (2.1), $\int_T \boldsymbol{i}^2 > 0$. Since $\boldsymbol{x}$ is an allocation, it follows that $\int_T \boldsymbol{x}^2 > 0$, so there must be a nonnull set of traders t for whom $x^2(t) > 0$. Now for any trader t, it follows from desirability (2.2) that

$$\boldsymbol{x}(t) + (1, 0, \ldots, 0) \succ_t \boldsymbol{x}(t).$$

Hence choosing t so that $\boldsymbol{x}^2(t) > 0$, we deduce from continuity (2.3) that for sufficiently small $\delta > 0$,

$$\boldsymbol{x}(t) + (1, -\delta, 0, \ldots, 0) \succ_t \boldsymbol{x}(t).$$

Then by (4.2),

$$\begin{aligned} p \cdot \boldsymbol{i}(t) &\leqq p \cdot [\boldsymbol{x}(t) + (1, -\delta, 0, \ldots, 0)] \\ &= p \cdot \boldsymbol{x}(t) + p^1 - \delta p^2 \\ &= p \cdot \boldsymbol{x}(t) - \delta p^2 \\ &< p \cdot \boldsymbol{x}(t), \end{aligned}$$

contradicting (4.3). This proves $p > 0$.

To demonstrate (4.4), let $x \in \boldsymbol{F}(t)$. Suppose first that $\boldsymbol{i}(t) \geq 0$; then $p \cdot \boldsymbol{i}(t) > 0$, because $p > 0$. So by (4.2), $p \cdot x > 0$; hence, there is j such that $x^j > 0$; let $j = 1$, say. From continuity (2.3), it then follows that $x - (\delta, 0, \ldots, 0) \in \boldsymbol{F}(t)$ for sufficiently small $\delta > 0$. Then by (4.2),

$$p \cdot \boldsymbol{i}(t) \leqq p \cdot [x - (\delta, 0, \ldots, 0)] = p \cdot x - \delta p^1 < p \cdot x,$$

proving (4.4). If $\boldsymbol{i}(t) = 0$ and $x \geq 0$, then clearly $p \cdot x > 0 = p \cdot \boldsymbol{i}(t)$, proving (4.4). Finally, suppose $\boldsymbol{i}(t) = 0$ and $x = 0$. Since $x \in \boldsymbol{F}(t)$, this means that $\boldsymbol{i}(t) = 0 \succ_t \boldsymbol{x}(t)$. If the set S of traders t for whom this happens is null, then it can be ignored; if, on the other hand, it has positive measure, then $\boldsymbol{i}$ dominates $\boldsymbol{x}$ via S, contradicting the membership of $\boldsymbol{x}$ in the core. This completes the proof of the main theorem.

3.5. Discussion of the Literature

The concept of competitive equilibrium is so well known in economics that we content ourselves with citing the papers of Walras [**23**], Wald [**22**], Arrow–Debreu [**1**], and McKenzie [**14**]; which are but a selection from an enormous literature. In discussing competitive equilibrium, Wald [**22**] acknowledged that the economic validity of his considerations is based on the assumption that "each of the participants is of the opinion that his own transactions do not influence the prevailing prices." But, like most workers in this area, he did not consider it necessary to modify either the finiteness of the model or the definition of competitive equilibrium. Von Neumann and Morgenstern saw more deeply into the problem: "The fact that every participant is influenced by the anticipated reactions of the others to his own measures... is most strikingly the crux of the matter in the classical problems of duopoly, oligopoly, etc. When the number of participants becomes really great, some hope emerges that the influence of every particular participant will become negligible.... These are, of course, the classical conditions of 'free competition' .... The current assertions concerning free competition appear to be very valuable surmises and inspiring anticipations of results. But they are not results, and it is scientifically unsound to treat them as such" [**16**, pp. 13–14]. The results of Scarf–Debreu

[**6, 7, 18**], Shubik [**21**], and this paper are presumably precisely the kind of "results" to which von Neumann and Morgenstern were referring.

The concept of core is well known in game theory. Though first named and intensively studied by Gillies and Shapley,[6] it had been used repeatedly already by von Neumann and Morgenstern [**16**]. This work was confined to games with side payments. An extension of the core notion to games without side payments — of which the market under consideration is an example — was made by Aumann and Peleg [**2, 3**]. The notion of Pareto optimality is related to that of core, but only distantly: Whereas the core consists of all those outcomes of a game with the property that no coalition of players can do better through its own efforts alone, Pareto optimality demands this only for the all-player coalition.

In economics, Edgeworth [**8**] was the first to investigate the core, albeit under a different name ("contract curve"). He, and subsequently Shubik [**21**], considered markets with two commodities and two "types" of traders, where traders of the same "type" have the same initial bundles and the same preferences. Under appropriate conditions,[7] they showed that, as the number of traders of each type tends to infinity, the core of the market in a certain sense "shrinks to a limiting point"; this point can be identified with the equilibrium allocation (which is unique in the markets they were considering). Recently, Scarf and Debreu [**7, 18**] generalized this work considerably by allowing an arbitrary but fixed finite number k of "types" of traders, rather than just two, and an arbitrary number of commodities. As the number of traders of each type tends to ∞, the core "shrinks to a limiting set," which can be identified with the set of equilibrium allocations. In this process, the number k of types is held fixed.

[6]The term "core" was introduced by Gillies and Shapley during a study of the properties of the von Neumann–Morgenstern solutions (see [**9, 10**]). The core as an independent solution concept was developed by Shapley in lectures at Princeton University in the fall of 1953.

[7]Which differ in the two investigations.

These results are special cases of the rough conjecture stated in the introduction that "the core approaches the set of equilibrium allocations as the number of traders tends to infinity." Supplying a precise statement of this is a nontrivial problem. The trouble is that the core and the set of equilibrium allocations are subsets of a space whose dimension varies with the number of traders. Thus as we add traders, the space under consideration changes, and in such a context it is not clear what "approach" means. It was because of this difficulty that Debreu and Scarf [**7**] postulated a fixed finite number of types of traders; together with some other assumptions, this enabled them to work in a fixed finite dimensional space. The other assumptions are that the preferences are quasi-orders[8] and that the indifference surfaces are strictly convex (i.e., convex and contain no straight line segment).

The notion of finitely many types might not at first sight seem objectionable. But it involves the further assumption that there are "many" traders of each type; in fact the number of traders of each type must be very large compared to the number of types in order for their model to be applicable. This seems far from economic reality, where, in general, different traders cannot be expected to have the same initial bundles or the same preferences. The continuous model allows *all* traders to have different initial bundles and different preferences.[9] There is no problem of working in spaces of varying dimension, because we start with a space of *infinite* dimension and remain in this same space throughout the investigation. For this reason also, it is not necessary to assume that the preferences are quasi-orders or that they are convex. Incidentally, Debreu and Scarf assume that each trader starts with a positive amount of each commodity, whereas we dispense with this assumption.

[8]A "quasi-order" is a transitive, reflexive, and complete relation. It is usually denoted $\succsim$, not $\gtrsim$, and is to be thought of intuitively as "preference-or-indifference."

[9]It does not, of course, insist on this; a coalition of positive measure — or even all the traders — may have the same initial bundles and preferences. The concept of "type" is simply irrelevant in our approach.

As we stated in the introduction, much of our proof is adapted from that of Debreu–Scarf. The key to the proof is Lemma 4.1. Debreu–Scarf prove the corresponding lemma by using the finiteness of the number of types. Our proof uses the same basic idea, after noting that the set of traders can be partitioned into a large number of nonnull coalitions whose members, though not identical, are fairly similar. The idea involved here is the same as in the approximation of an arbitrary function by a simple function, i.e., a function taking finitely many values.

Scarf [**18**] and Debreu [**6**] have also investigated a model with a *denumerable* infinity of traders. The assumptions are similar to those of [**7**], including finitely many types, positivity of initial bundles, and preferences are quasi-orders and are convex. Other references for games with denumerably many players are Kalisch and Nering [**11**] and Shapley [**20**], but these are in a non-economic context.

There are two aspects of the continuous model that we have not discussed. The first is the question of the existence of a competitive equilibrium, or equivalently, the nonemptiness of the core. We have proved only that the core equals the set of equilibrium allocations, but it may well happen that both are empty. It is possible to prove an existence theorem if it is assumed that the preferences are quasi-orders[8] (but it is not necessary to assume convexity). We plan to publish details in a subsequent paper.

The other aspect we have not discussed is the economic significance of the Lebesgue measure of a coalition. This, too, we plan to discuss in a subsequent paper.

It is a pleasure to acknowledge inspiration received from the work of Debreu, Scarf, and Shapley, and from a stimulating correspondence with Professor Debreu. Also, we thank Mr. A. Shlein for an enlightening conversation.

References

1. Arrow, K. J., and G. Debreu: "Existence of an Equilibrium for a Competitive Economy," *Econometrica*, Vol. 22 (1954), pp. 265–90.
2. Aumann, R. J., and B. Peleg: "Von Neumann-Morgenstern Solutions to Cooperative Games without Side Payments," *Bulletin of the American Mathematical Society*, Vol. 66 (1960), pp. 173–79.

3. Aumann, R. J.: "The Core of a Cooperative Game without Side Payments," *Transactions of the American Mathematical Society*, Vol. 98 (1961), pp. 539–52.
4. Davis, M.: "Symmetric Solutions to Symmetric Games with a Continuum of Players," in *Recent Advances in Game Theory* **[13]**, pp. 119–26.
5. Debreu, G.: *Theory of Value*, John Wiley, 1959.
6. ———: "On a Theorem of Scarf," *Review of Economic Studies*, June, 1963.
7. Debreu, G., and H. Scarf: "A Limit Theorem on the Core of an Economy," *International Economic Review*, September, 1963.
8. Edgeworth, F. Y.: *Mathematical Psychics,* London, 1881.
9. Gillies, D. B.: *Some Theorems on n-Person Games*, Ph.D. Thesis, Department of Mathematics, Princeton University, 1953.
10. ———: "Solutions to General Non-Zero Sum Games," in *Contributions to the Theory of Games IV* **[12]**, pp. 47–85.
11. Kalisch, G. K., and E. D. Nering: "Countably Infinitely Many Person Games," in *Contributions to the Theory of Games IV* **[12]**, pp. 43–45.
12. Luce, R. D., and A. W. Tucker (editors): *Contributions to the Theory of Games IV*, Annals of Mathematics Studies No. 40, Princeton University Press, 1959.
13. Maschler, M. (editor): *Recent Advances in Game Theory*, Proceedings of the Princeton University Conference of October, 1961, privately printed for members of the conference.
14. McKenzie, L. W.: "On the Existence of General Equilibrium for a Competitive Market," *Econometrica*, Vol. 27 (1959), pp. 54–71.
15. Milnor, J. W., and L. S. Shapley: "Values of Large Games II: Oceanic Games," *Mathematics of Operations Research*, Vol. 3 (1978), pp. 290–307.
16. Neumann, J. von, and O. Morgenstern: *Theory of Games and Economic Behavior*, Princeton University Press, 1944 (Third edition, 1953).
17. Peleg, B.: "Quota Games with a Continuum of Players," *Israel Journal of Mathematics*, 1 (1963), pp. 48–53.
18. Scarf, H.: "An Analysis of Markets with a Large Number of Participants," in *Recent Advances in Game Theory* **[13]**, pp. 127–55.
19. Shapley, L. S.: "Values of Large Games III: A Corporation with Two Large Stockholders," *The RAND Corporation*, RM 2650 PR, December, 1961 (multilithed).
20. ———: "Values of Games with Infinitely Many Players," in *Recent Advances in Game Theory* **[13]**, pp. 113–118.
21. Shubik, M.: "Edgeworth Market Games," in *Contributions to the Theory of Games IV* **[12]**, pp. 267–78.
22. Wald, A.: "Ueber einige Gleichungssysteme der mathematischen Ökonomie," *Zeitschrift für Nationalökonomie*, Vol. 7 (1936), pp. 637–70 (translated as "On Some Systems of Equations of Mathematical Economics," *Econometrica*, Vol. 19 (1951) pp. 368–403).
23. Walras, L.: *Mathematische Theorie der Preisbestimmung der wirtschaftlichen Güter*, Stuttgart, Ferdinand Enke, 1881.

Chapter 4

Subjectivity and Correlation in Randomized Strategies*

4.1. Introduction

Subjectivity and correlation, though formally related, are conceptually distinct and independent issues. We start by discussing subjectivity.

A mixed strategy in a game involves the selection of a pure strategy by means of a random device. It has usually been assumed that the random device is a coin flip, the spin of a roulette wheel, or something similar; in brief, an "objective" device, one for which everybody agrees on the numerical values of the probabilities involved. Rather oddly, in spite of the long history of the theory of subjective probability, nobody seems to have examined the consequences of basing mixed strategies on "subjective" random devices, i.e. devices on the probabilities of whose outcomes people may disagree (such as horse races, elections, etc.). Even a fairly superficial such examination yields some startling results, as follows:

a. Two-person zero-sum games lose their "strictly competitive" character. It becomes worthwhile to cooperate in such games, i.e. to enter into binding agreements.[1] The concept of the "value" of a zero-sum game loses some of its force, since *both* players can get more than the value (in the utility sense).

*Reprinted from *Journal of Mathematical Economics* (1974), 1, 67–96, with permission from Elsevier.

[1]Example 2.1 and Section 4.6.

b. In certain n-person games with $n \geqq 3$ new equilibrium points appear, whose payoffs strictly dominate the payoffs of all other equilibrium points.[2]

Result (a) holds not just for certain selected 2-person 0-sum games, but for practically[3] all such games. Moreover, it holds if there is any area whatsoever of subjective disagreement between the players, i.e., any event in the world (possibly entirely unconnected with the game under consideration) for which players 1 and 2 have different subjective probabilities.

The phenomenon enunciated in Result (b) shows that not only the 2-person 0-sum theory, but also the non-cooperative n-person theory is modified in an essential fashion by the introduction of this new kind of strategy. However, this phenomenon cannot occur[4] for 2-person games (zero-sum or not); for such games we will show[5] that the set of equilibrium payoff vectors is not changed by the introduction of subjectively mixed strategies.

We now turn to correlation. Correlated strategies are familiar from cooperative game theory, but their applications in non-cooperative games are less understood. It has been known for some time that by the use of correlated strategies in a non-cooperative game, one can achieve as an equilibrium any payoff vector in the convex hull of the mixed strategy (Nash) equilibrium payoff vectors. Here we will show that by appropriate methods of correlation, even points outside of this convex hull can be achieved.[6]

In describing these phenomena, it is best to view a randomized strategy as a random variable with values in the pure strategy space, rather than as a distribution over pure strategies. In Section 4.3 we develop such a framework; it allows for subjectivity, correlation, and

---

[2]Example 2.3.

[3]Specifically, wherever there is one payoff greater than the value and another one less than the value (for Player 1, say).

[4]Except in a very degenerate sense; see Example 2.2.

[5]Section 4.5.

[6]Examples 2.5 through 2.9.

all possible combinations thereof. Thus, a side product of our study is a descriptive theory (or taxonomy) of randomized strategies.

We are very grateful to Professors M. Maschler, B. Peleg, and R. W. Rosenthal for extremely stimulating correspondence and comments on this study.

4.2. Examples

Example 2.1. Consider the familiar two-person zero-sum game "matching pennies," which we write in matrix form as follows:

1, −1	−1, 1
−1, 1	1, −1

.

Let D be an event to which players 1 and 2 ascribe subjective probability 2/3 and 1/3 respectively. Suppose now that the players bindingly agree to the following pair of strategies: Player 2 will play left in any event; Player 1 will play top if D occurs, and bottom otherwise. The expectation of both players is then 1/3, whereas the value of the game is of course 0 to both players.

To make this example work, one needs an event D whose subjective probability is $>1/2$ for one player and $<1/2$ for the other. Such an event can always be constructed as long as the players disagree about something, i.e., there is some event B with $p_1(B) \neq p_2(B)$ (where p_i is the subjective probability of i). For example, if $p_1(B) = 7/8$ and $p_2(B) = 5/8$, then we can construct the desired event by tossing a fair coin twice (provided these particular coin tosses were not involved in the description of B); if C is the event "at least one head," then we have $p_1(B \cap C) = 21/32 > 1/2 > 15/32 = p_2(B \cap C)$. This kind of construction can always be carried out (see the proof of Proposition 5.1).

Example 2.2. Suppose that there is an event D to which players 1 and 2 ascribe subjective probabilities 1 and 0 respectively and such that only player 1 is informed before the play as to whether or not D occurs. Consider the following pair of strategies: Player 2 plays left; Player 1 plays top if D occurs, bottom otherwise. This strategy yields a payoff of (1, 1); moreover, it is actually in equilibrium, in the sense that neither player has any incentive to carry out a unilateral change in strategy. However, it has a somewhat degenerate flavor, since it requires that at least one of the two players be certain of a falsehood (see Section 4.8). We now show that a similar phenomenon can also occur in a non-degenerate set-up; but it requires 3 players.

Example 2.3. Consider the three-person game given as follows:

0, 8, 0	3, 3, 3
1, 1, 1	0, 0, 0

0, 0, 0	3, 3, 3
1, 1, 1	8, 0, 0

here Player 1 picks the row, Player 2 the column, and Player 3 the matrix. A triple of objectively mixed strategies is an equilibrium point in this game if and only if Player 1 plays bottom and Player 2 plays left; Player 3 may play any mixed strategy. All these equilibrium points have the same payoff vector, namely (1, 1, 1). Suppose now that D is an event to which players 1 and 2 ascribe subjective probabilities 3/4 and 1/4 respectively, and such that only Player 3 is informed as to whether or not D occurs. Consider the following strategy triple: Player 1 plays top; Player 2 plays right; Player 3 chooses the left matrix if D occurs, the right matrix if not. If Player 1 moves down he will get $1/4 \cdot 8 = 2$ rather than $1 \cdot 3 = 3$; similarly for Player 2 if he moves left; as for Player 3, he certainly cannot profit from moving, since he is getting his maximum payoff in the whole game. Therefore this is an equilibrium point; its payoff is (3, 3, 3), instead of the (1, 1, 1) obtained at all objective equilibrium points.

That this kind of phenomenon cannot occur with fewer than 3 players will be shown in Section 4.5.

An interesting feature of this example is that the higher payoff at the new, subjective, equilibrium point is not only "subjectively higher," it is "objectively higher." That is, unlike the case in Example 2.1, the payoff is higher not only because of the differing probability estimates of the players, but it is higher in any case, whether or not the event D takes place. The contribution of subjective probabilities in this case is to make the new point an equilibrium; once chosen, it is *sure* to make all the players better off than at any of the old equilibrium points.

We remark that by using different numbers we can find similar examples based on arbitrarily small probability differences.

It is essential in this example that only Player 3 be informed as to whether or not D occurs. If, say, Player 1 were also informed (before the time comes for choosing his pure strategy), he could do better by playing top or bottom according to whether D occurs on not.[7] This secrecy regarding D is quite natural. If we were to insist that all players be informed about D, it would be like insisting that in an objectively mixed strategy based on a coin toss, all players be informed as to the outcome of the toss. But much of the effectiveness of mixed strategies is based precisely on the secrecy, which would then be destroyed. In practical situations, it is of course quite common for some players to have differing subjective probabilities for events about which they are not informed, and on which other players peg their strategy choices.

Our last 6 examples deal with correlation.

Example 2.4. Consider the following familiar 2-person non-zero-sum game:

2, 1	0, 0
0, 0	1, 2

.

There are exactly 3 Nash equilibrium points: 2 in pure strategies, yielding (2, 1) and (1, 2) respectively, and one in mixed strategies, yielding (2/3, 2/3). The payoff vector (3/2, 3/2) is not achievable at all in (objectively) mixed strategies. It is, however, achievable in "correlated" strategies, as follows: One fair coin is tossed; if it falls heads, players 1 and 2 play top and left respectively; otherwise, they play bottom and right.

The interesting aspect of this procedure is that it not only achieves (3/2, 3/2), it is also in equilibrium; neither player can gain

[7] I am grateful to R. W. Rosenthal for this remark.

by a unilateral change. Any point in the convex hull of the Nash equilibrium payoffs of any game can be achieved in a similar fashion, and will also be in equilibrium. This is not new; it has been in the folklore of game theory for years. I believe the first to notice this phenomenon (at least in print) were Harsanyi and Selten (1972).

In the following 4 examples we wish to point out a phenomenon that we believe *is* new, namely that by the use of correlated strategies one can achieve a payoff vector that is in equilibrium in the same sense as above, but that is *outside* the convex hull of the Nash equilibrium payoffs. In fact, except in Example 2.7, it is better for all players than any Nash equilibrium payoff.

Example 2.5. Consider the 3-person game given as follows:

0, 0, 3	0, 0, 0
1, 0, 0	0, 0, 0

2, 2, 2	0, 0, 0
0, 0, 0	2, 2, 2

0, 0, 0	0, 0, 0
0, 1, 0	0, 0, 3

.

Here Player 1 picks the row, Player 2 the column, and Player 3 the matrix. If we restrict ourselves to pure strategies, there are only 3 equilibrium payoffs, namely (0, 0, 0), (1, 0, 0), and (0, 1, 0). If we allow (objectively) mixed strategies, some more equilibrium payoffs are added, but none of their coordinates exceed 1. Consider now the following strategy triple: Player 3 plays the middle matrix. Players 1 and 2 get together and toss a fair coin, but do not inform Player 3 of the outcome of the toss. If the coin falls on heads, players 1 and 2 play top and left respectively; otherwise, they play bottom and right respectively. The payoff is (2, 2, 2). If Player 3 would know the outcome of the toss, he would be tempted to move away; for example, if it was heads, he would move left. Since he does not know, he would lose by moving. Thus the introduction of correlation among *subsets* of the players can significantly improve the payoff to everybody.

Example 2.6. Another version of Example 2.5 is the following:

0, 1, 3	0, 0, 0
1, 1, 1	1, 0, 0

2, 2, 2	0, 0, 0
2, 2, 0	2, 2, 2

0, 1, 0	0, 0, 0
1, 1, 1	1, 0, 3

.

This version has the advantage that there is only one Nash equilibrium payoff, namely (1, 1, 1); and it can be read off from the matrices by the use of simple domination arguments, as compared to the slightly laborious computations needed in the previous example. The advantage of Example 2.5 is that the new correlated strategy equilibrium point has the property that any deviation will actually lead to a loss (not only a failure to gain).

In both examples, Player 3 will not even *want* to know the outcome of the toss; he will *want* players 1 and 2 to perform it in secret. It is important for him that players 1 and 2 know that he does not know the outcome of the toss; otherwise they cannot depend on him to choose the middle matrix, and will in consequence themselves play for an equilibrium point that is less advantageous for all. Thus in Example 2.6, Player 3 knows that if he can "peep," then players 1 and 2 will necessarily play bottom and left respectively, to the mutual disadvantage of all.

Example 2.7. In Examples 2.5 and 2.6, equilibrium payoffs outside of the convex hull of the Nash equilibrium payoffs were achieved by "partial" correlation — correlation of strategy choices by 2 out of the 3 players. We now show that a similar phenomenon can occur even in 2-person games; here again a kind of partial correlation is used, but it is subtler than that appearing previously.

Consider the 2-person game given as follows:

6, 6	2, 7
7, 2	0, 0

.

This game has two pure strategies equilibrium points, with payoff (2, 7) and (7, 2) respectively, and one mixed strategy equilibrium point, with payoff $(4\frac{2}{3},\ 4\frac{2}{3})$. Consider now an objective chance mechanism that chooses one of three points A, B, C with probability $1/3$ each. After the point has been chosen, Player 1 is told whether or not A was chosen, and Player 2 is told whether or not C was chosen; nothing more is told to the players. Consider the following pair of strategies: If he is informed that A was chosen, Player 1 plays his bottom strategy;

otherwise, his top strategy. Similarly, if he is informed that C was chosen, Player 2 plays his right strategy; otherwise, he goes left. It is easy to verify that this pair of strategies is indeed in equilibrium, and that it yields the players an expectation of (5, 5) — a payoff that is outside the convex hull of the Nash equilibrium payoffs (listed above).

The strategies in this equilibrium point are not stochastically independent, like the mixed or pure strategies appearing in Nash equilibrium points, nor are they totally correlated, like the strategies appearing in Example 2.4. It is precisely the partial correlation that enables the phenomenon we observe here.

Example 2.8. Consider the 2-person game given as follows:

6, 6	0, 0	2, 7
0, 0	4, 4	3, 0
7, 2	0, 3	0, 0

.

This is obtained from the previous example by adding a middle row and a middle column, with appropriate payoffs. The strategy pairs (top, right) and (bottom, left), which previously yielded the equilibrium payoffs (2, 7) and (7, 2), are no longer in equilibrium; the equilibrium property has been "killed" by the addition of the new strategies. The mixed-strategy pair which previously yielded $(4\frac{2}{3}, 4\frac{2}{3})$ remains in equilibrium here; moreover we get a new pure strategy equilibrium point yielding (4, 4), and a new mixed strategy equilibrium point yielding $(2\frac{10}{23}, 2\frac{10}{23})$. These are all the Nash equilibrium payoffs. Thus we see that $4\frac{2}{3}$ is the maximum that either player can get at any Nash equilibrium point.

The equilibrium point in "partially correlated" strategies described in Example 2.7 remains in equilibrium here, and yields (5, 5) — more, for both players, than that yielded by any of the Nash equilibria.

Though the random device needed in Examples 2.7 and 2.8 is conceptually somewhat more complex than the coin tosses (secret or joint) used in classical game theory and in Examples 2.4, 2.5, and 2.6, it is not at all difficult to construct. Given a roulette wheel, it is easy to attach electrical connections that will do the job. If the reader wishes, he can think of the players as jointly supervising the construction of the device, and then retiring to separate rooms to get the information and choose their strategies. It is advantageous for both players to build the device, to satisfy each other that it is working properly, and then to follow the above procedure. Once chosen, the procedure is of course self-enforcing; neither player will wish to renege at any stage.

Example 2.9. Consider again the 2-person 0-sum game "matching pennies" already treated in Example 2.1. Suppose that nature chooses one of four points A, B, C, or D, and that players 1 and 2 ascribe to these choices subjective probabilities (1/3, 1/6, 1/6, 1/3) and (1/6, 1/3, 1/3, 1/6), respectively. Under no condition is either player told which point was chosen, but Player 1 is told whether the point chosen is in $\{A, B\}$ or in $\{C, D\}$, and Player 2 is told whether the point chosen is in $\{A, C\}$ or in $\{B, D\}$. Consider the following strategy pair: Player 1 chooses top or bottom according to whether he is told "$\{A, B\}$" or "$\{C, D\}$"; Player 2 chooses left or right according to whether he is told "$\{A, C\}$" or "$\{B, D\}$." Like the strategy pair in Example 2.1, this yields each player an expectation of 1/3, whereas the value of the game is of course 0 to both players; unlike the strategy pair of Example 2.1, *this pair is in equilibrium.*

If, in a 2-person zero-sum game with value v, we permit correlation but rule out subjectivity, then the payoffs to all strategy pairs continue to sum to 0, and it is easy to see that the only equilibrium points have payoffs $(v, -v)$. If we rule out correlation (i.e. permit mixed or pure strategies only) but permit subjectivity, we still get only $(v, -v)$ as an equilibrium payoff; this follows from Proposition 5.1. But if we permit both subjectivity and correlation, then this example shows that there may exist mutually advantageous equilibrium points, even in 2-person 0-sum games.

4.3. The Formal Model

In this section, definitions are indicated by italics.

A *game* consists of:

1. a finite set N (the *players*); write $N = \{1, \ldots, n\}$;
2. for each $i \in N$, a finite set S_i (the *pure strategies* of i);
3. a finite set X (the *outcomes*);
4. a function g from the cartesian product $S = \boldsymbol{x}_{i \in n} S_i$ onto X (*the outcome function*).

This completes the description of the game as such; formally, however, we still need some equipment for randomizing strategies, and for defining utilities and subjective probabilities for the players. Thus to the description of the game we append the following:

5. A set Ω (the *states of the world*), together with a σ-field $\mathscr{B}$ of subsets of Ω (the *events*);
6. For each player i, a sub-σ-field $\mathscr{I}_i$ of $\mathscr{B}$ (the events in $\mathscr{I}_i$ are those *regarding which i is informed*).
7. For each player i, a relation $\succsim_i$ (the *preference order of i*) on the space of lotteries on the outcome space X, where a *lottery* on X is a $\mathscr{B}$-measurable[8] function from Ω to X.

The intuitive scenario associated with this model involves the following steps, to be thought of as occurring one after the other as follows:

i. Nature chooses a point ω in Ω.
ii. Each player i is informed as to which events in $\mathscr{I}_i$ contain ω.
iii. Each player i chooses[9] a pure strategy s_i in S_i.
iv. The outcome is determined[10] to be $g(s_1, \ldots, s_n)$.

---

[8]This means that for each $x \in X$, the set $\{\boldsymbol{x} = x\}$ [i.e. the set $\{\omega \colon \boldsymbol{x}(\omega) = x\}$] is in $\mathscr{B}$.

[9]Without informing the other players.

[10]Note that g does not depend on ω; see subsection (d) of Section 4.9.

Returning to the formal model, let us define a *randomized strategy* (or simply *strategy*) for player i to be a measurable function $\boldsymbol{s}_i$ from $(\Omega, \mathscr{I}_i)$ to S_i. Note that $\boldsymbol{s}_i$ must be measurable w.r.t. (with respect to) $\mathscr{I}_i$, not merely w.r.t. $\mathscr{B}$; this means that i can peg his strategies only on events regarding which he is informed. [In the above scenario, strategies are chosen before step (ii).]

If $\boldsymbol{s}$ is an n-tuple of strategies, then $g(\boldsymbol{s})$ is a lottery on X. An *equilibrium point* is an n-tuple $\boldsymbol{s} = (\boldsymbol{s}_1, \ldots, \boldsymbol{s}_n)$ of strategies such that for all i and all strategies $\boldsymbol{t}_i$ of i, we have

$$g(\boldsymbol{s}) \succsim_i g(\boldsymbol{s}_1, \ldots, \boldsymbol{s}_{i-1}, \boldsymbol{t}_i, \boldsymbol{s}_{i+1}, \ldots, \boldsymbol{s}_n).$$

We will assume:

Assumption I. For each i, there is a real function u_i on X (the *utility function* of i) and a probability measure[11] p_i on Ω (the *subjective probability* of i) such that for all lotteries $\boldsymbol{x}$ and $\boldsymbol{y}$ on X, we have

$$\boldsymbol{x} \succsim_i \boldsymbol{y} \text{ if and only if}$$
$$\int_\Omega u_i(\boldsymbol{x}(\omega))\mathrm{d}p_i(\omega) \geqq \int_\Omega u_i(\boldsymbol{y}(\omega))\mathrm{d}p_i(\omega);$$

moreover, p_i is unique.[12]

This assumption says that player i's preferences between lotteries are governed by the expected utility of the outcomes of the lotteries. There are well-known systems of axioms on the preferences $\succsim_i$ that lead to utilities and subjective probabilities; see for example Savage (1954) or Anscombe and Aumann (1963).

[11] σ-additive non-negative measure with $p_i(\Omega) = 1$.

[12] I.e. if (u_i', p_i') also satisfy the condition of the previous sentence, then $p_i' = p_i$. Axiom systems leading to subjective probabilities usually imply the uniqueness as well. In our case, if we had wanted to minimize our assumptions we could have avoided assuming uniqueness, at this stage. Indeed, the uniqueness of p_i follows from the *existence* of a non-atomic p_i; and this is assumed in Assumption II. It is, however, more convenient to assume uniqueness already at this stage, since it enables us to refer immediately to "the" subjective probability of player i, and this simplifies the entire discussion.

Note that p_i is defined on all of $\mathscr{B}$, not only on $\mathscr{I}_i$; that is, i assigns a probability to all events, not only those regarding which i is informed.

Two events A and B are *i-independent* if

$$p_i(A \cap B) = p_i(A)p_i(B);$$

intuitively, this means that i can get no hint about A from information regarding B. More generally, the events $A_1, \ldots, A_k$ are *i-independent* if

$$p_i(B_1 \cap \cdots \cap B_k) = p_i(B_1) \cdots p_i(B_k)$$

whenever each B_j is either A_j or Ω. Events are *independent* if they are i-independent for all i. An event is *i-secret* if it is in $\mathscr{I}_i$, and for each j other than i, it is j-independent of all events in the σ-field generated by all the $\mathscr{I}_k$ with $k \neq i$. Thus i is informed regarding each i-secret event, but players other than i can get no hint of it, even by pooling their knowledge. The family of i-secret events is denoted $\mathscr{S}_i$. We assume

Assumption II. For each i, there is a σ-field $\mathscr{R}_i$ of i-secret events such that each p_j is non-atomic on $\mathscr{R}_i$.

Intuitively, $\mathscr{R}_i$ can be constructed from a roulette spin conducted by i in secret. To obtain the non-atomicity, it is sufficient to assume that each player j assigns subjective probability 0 to each particular outcome;[13] or equivalently, that there exist finite partitions of Ω into $\mathscr{R}_i$-events whose p_j-probabilities are arbitrarily small. Such non-atomicity assumptions are familiar in treatments of subjective probability.[14] Note that the roulette wheel need not be "objective," i.e. the players may disagree about the probabilities involved.

A strategy $\boldsymbol{s}_i$ of i is *mixed* if it is $\mathscr{S}_i$-measurable, i.e. pegged on i-secret events. Strategies $\boldsymbol{s}_1, \ldots, \boldsymbol{s}_n$ are *uncorrelated* if for each

[13]The set of outcomes is taken to be the unit interval (*not* $\{0, 1, \ldots, 36\}$).
[14]Cf. Savage (1954, pp. 38–40, especially postulates P6′ and P6).

$s \in S$, the n events $\{\boldsymbol{s}_j = s_j\}$ are independent. Mixed strategies are uncorrelated, but the converse is false; strategies may be uncorrelated even though they are pegged on events regarding which all players are informed.

An event A is called *objective* if all the subjective probabilities $p_i(A)$ coincide; in that case their common value is called the *probability* of A, and is denoted $p(A)$. A strategy $\boldsymbol{s}_i$ of i is called *objective* if it is pegged on objective events, i.e. if $\{\boldsymbol{s}_i = s_i\}$ is an objective event for all $s_i \in S_i$. The strategies occurring in the classical non-cooperative theory are precisely those that are both objective and mixed.

An event or strategy is called *subjective* if it is not objective.

The preference order of i determines his utility function on X only up to a monotonic linear transformation. Given a specific choice $u_1, \ldots, u_n$ of utility functions of the players, define the *payoff function* of player i in the usual way; that is, if s is an n-tuple of pure strategies (i.e. $s \in S$), define

$$h_i(s) = u_i(g(s));$$

and if $\boldsymbol{s}$ is an n-tuple of arbitrary strategies, define

$$H_i(\boldsymbol{s}) = E_i(h_i(\boldsymbol{s})) = \int_\Omega h_i(\boldsymbol{s}(\omega))\mathrm{d}p_i(\omega), \tag{3.1}$$

E_i being the expectation operator w.r.t. the probability measure p_i on Ω. Then $g(\boldsymbol{s}) \succ_i g(\boldsymbol{t})$ if and only if $H_i(\boldsymbol{s}) > H_i(\boldsymbol{t})$. Write

$$h = (h_1, \ldots, h_n),$$
$$H = (H_1, \ldots, H_n).$$

Any vector of the form $H(\boldsymbol{s})$ is called a *feasible payoff*; if $\boldsymbol{s}$ is an equilibrium point, then $H(\boldsymbol{s})$ is called an *equilibrium payoff*; and if $\boldsymbol{s}$ is an equilibrium point in objective mixed strategies, then $H(\boldsymbol{s})$ is an *objective mixed equilibrium payoff.*

To orient the reader, we mention that Example 2.1 involves a pair of subjective strategies that is not in equilibrium; Examples 2.2 and 2.3 involve equilibrium points in mixed subjective strategies;

and Examples 2.4 through 2.8 involve equilibrium points that are objective, but are not in mixed strategies (the strategies are in fact correlated, i.e. not uncorrelated).

4.4. Preliminaries and Generalities

This section is devoted to the *statement* of several lemmas needed in the sequel, of an existence theorem for equilibrium points (Proposition 4.3), and of a proposition concerning the nature of the sets of feasible and equilibrium payoffs. We also discuss the notion of "correlation."

Though most of the results stated in this section are intuitively unsurprising, the proofs are a little involved (because of the relatively weak assumptions we have made). *We therefore postpone these proofs to Section 4.7*, in order to get as quickly as possible to the conceptually more interesting results of the paper (Propositions 5.1 and 6.1). Readers who are interested only in the statements of these propositions (as opposed to the proofs) may skip this section.

The first lemma asserts that objective mixed strategies can be constructed with arbitrary probabilities, i.e. that all the "classical" mixed strategies appear in this model as well. Define a *distribution* on S_i to be a real-valued function on S_i whose values are non-negative and sum to 1.

Lemma 4.1. *Let $i \in N$, and let σ_i be a distribution on S_i. Then i has an objective mixed strategy $\boldsymbol{s}_i$ such that for all $s_i \in S_i$,*

$$p\{\boldsymbol{s}_i = s_i\} = \sigma_i(s_i).$$

Next, we have

Lemma 4.2. *For any i in N, any event B, and any α between 0 and 1, there is an objective i-secret event with probability α that is independent of B.*

Intuitively, Lemmas 4.1 and 4.2 depend on the existence of an *objective* roulette wheel that i can spin in secret. Thus they appear to go somewhat further[15] than Assumption II, and it is of some interest that in fact, they follow from it.

The next proposition asserts that the classical equilibrium points of Nash (1951) appear in this model as well. If $\sigma = (\sigma_1, \ldots, \sigma_n)$ is an n-tuple of distributions on $S_1, \ldots, S_n$ respectively, and if $i \in N$, define

$$F_i(\sigma) = \Sigma_s h_i(s) \Pi_{j \in N^{\alpha_j}}(s_j),$$

where the sum runs over all pure strategy n-tuples $s = (s_1, \ldots, s_n)$. The *payoff to* σ is the n-tuple $F(\sigma) = (F_1(\sigma), \ldots, F_n(\sigma))$. Recall that σ is a *Nash equilibrium point* if for any i and any distribution τ_i on S_i, we have

$$F_i(\sigma) \geqq F_i(\sigma_1, \ldots, \sigma_{i-1}, \sigma_{i+1}, \ldots, \sigma_n);$$

in that case $F(\sigma)$ is called a *Nash equilibrium payoff.*

Proposition 4.3. *The set of Nash equilibrium payoffs coincides with the set of objective mixed equilibrium payoffs.*

From this proposition and Nash's theorem, it follows immediately that there is an equilibrium point in every game.

We now come to the concept of correlation. Set $S_N = \bigcap_{i \in N} \mathscr{I}_i$; the members of S_N are called *public events.* A *continuous chance device,* or *roulette* for short, is a sub-σ-field $\mathscr{R}$ of Ω on which each p_j is non-atomic; if $\mathscr{R}$ consists of public events, it is called a *public roulette.*

Parallel to Lemma 4.2, we have

Lemma 4.4. *Assume that there is a public roulette. Then for any event B, and any α between* 0 *and* 1, *there is an objective public event with probability α that is independent of B.*

A public roulette can be used as a correlating device on which all players can peg their choices; this leads (cf. Example 2.4) to

[15] See subsection (e) of Section 4.9 for a discussion of this point.

Proposition 4.5. *Assume that there is a public roulette. Then the set of equilibrium payoffs and the set of feasible payoffs are both convex.*

It is not known whether the set of equilibrium — or feasible — payoffs is closed, whether or not one assumes the existence of a public roulette.

Public roulettes enable *all* players to correlate their strategies, as in Example 2.4. In Examples 2.5 through 2.9, the correlation is of a subtler kind. To describe the situation, let us call the triple consisting of the pair $(\Omega, \mathscr{B})$, the n-tuple $(\mathscr{I}_1, \ldots, \mathscr{I}_n)$, and the n-tuple $(p_1, \ldots, p_n)$ a *randomizing structure.* In Examples 2.4, 2.5, and 2.6, the randomizing structure is of a particularly simple kind, which we call *standard*, and which is described as follows:

For each $T \subset N$, let J_T be a copy of the unit interval [0, 1] with the Borel sets. Let $(\Omega, \mathscr{B}) = \mathbf{x}_{T \subset N} J_T$, and let π_T be the projection of Ω on J_T. For i in N, call two members ω_1 and ω_2 of Ω *i-equivalent* if $\pi_T(\omega_1) = \pi_T(\omega_2)$ for all T containing i. Let $\mathscr{I}_i$ be the σ-field of all events in Ω that are unions of i-equivalence classes. Let all the p_i be Lebesgue measure on Ω.

Intuitively, the points of J_T are outcomes of an objective roulette spin conducted in the presence of the members of T only. Note that all the probabilities are the same, so that subjectivity does not enter the picture. Even in this relatively simple case, though, it is not clear that the set of equilibrium payoffs is closed (it is convex by Proposition 4.5). However, when $n = 2$ and the randomizing structure is standard, then it is easily verified that the set of equilibrium payoffs is precisely the convex hull of the Nash equilibrium payoffs.

In Example 2.7, the randomizing structure is not standard; but the probabilities are objective, i.e. all the p_i coincide. It is easily verified that whenever the probabilities are objective and there is a public roulette, the set of feasible payoffs is simply the convex hull of the pure strategy payoffs. In particular, this will be the case when the randomizing structure is standard.

4.5. Equilibrium Points in Two-Person Games

In Example 2.3 we showed that by pegging strategies on subjective events, it is possible to find a 3-person game with an equilibrium point in mixed strategies whose payoff is higher for *all* players than the payoff to any equilibrium in objective mixed strategies. In this section we show that this cannot happen in the case of 2-person games.

Proposition 5.1. *Let G be a 2-person game,*[16] *and assume that*

$$\textit{for any event } B, p_1(B) = 0 \textit{ if and only if } p_2(B) = 0. \tag{5.2}$$

Then for each equilibrium point $\boldsymbol{s}$ in mixed strategies, there is an equilibrium point $\boldsymbol{t}$ in objective mixed strategies such that $H(\boldsymbol{s}) = H(\boldsymbol{t})$.

Proof. Let $(\boldsymbol{s}_1, \boldsymbol{s}_2)$ be an equilibrium point in mixed strategies. Let $\boldsymbol{t}_1$ be an objective mixed strategy for Player 1 that mimics the way Player 2 sees $\boldsymbol{s}_1$, i.e., such that for all pure strategies s_1 of Player 1,

$$p\{\boldsymbol{t}_1 = s_1\} = p_2\{\boldsymbol{s}_1 = s_1\}; \tag{5.3}$$

such a strategy exists because of Lemma 4.1. Similarly, let $\boldsymbol{t}_2$ be an objective mixed strategy for Player 2 such that for all s_2,

$$p\{\boldsymbol{t}_2 = s_2\} = p_1\{\boldsymbol{s}_2 = s_2\}. \tag{5.4}$$

Let $s_{11}, s_{12}, \ldots$ be the pure strategies that enter actively into $\boldsymbol{s}_1$, i.e., with positive p_1-probability. Because $\boldsymbol{s}_1$ is in equilibrium,

$$E_1 h_1(s_{1j}, \boldsymbol{s}_2) = H_1(\boldsymbol{s}_1, \boldsymbol{s}_2) \tag{5.5}$$

for all j. But from (5.2) it follows that the s_{1j} are also precisely the pure strategies that enter actively into $\boldsymbol{t}_1$; hence (5.5) yields

$$H_1(\boldsymbol{t}_1, \boldsymbol{s}_2) = H_1(\boldsymbol{s}_1, \boldsymbol{s}_2). \tag{5.6}$$

[16] Not necessarily 0-sum.

Since $\boldsymbol{s}_1$ maximizes H_1 (if 2 plays $\boldsymbol{s}_2$), it follows from (5.6) that $\boldsymbol{t}_1$ also does. But by (5.4), Player 1 ascribes the same distribution to $\boldsymbol{t}_2$ as to $\boldsymbol{s}_2$; therefore

$$H_1(\boldsymbol{t}_1, \boldsymbol{t}_2) = H_1(\boldsymbol{t}_1, \boldsymbol{s}_2) = H_1(\boldsymbol{s}_1, \boldsymbol{s}_2),$$

and furthermore, $\boldsymbol{t}_1$ maximizes H_1 if 2 plays $\boldsymbol{t}_2$. Similarly,

$$H_2(\boldsymbol{t}_1, \boldsymbol{t}_2) = H_2(\boldsymbol{s}_1, \boldsymbol{t}_2) = H_2(\boldsymbol{s}_1, \boldsymbol{s}_2),$$

and $\boldsymbol{t}_2$ maximizes H_2 if 1 plays $\boldsymbol{t}_1$. Therefore $(\boldsymbol{t}_1, \boldsymbol{t}_2)$ is an objective mixed equilibrium point with the same payoff as $(\boldsymbol{s}_1, \boldsymbol{s}_2)$, and the proof of this proposition is complete.

It would have been reasonable to conjecture that this proposition remains true if both occurrences of the word "mixed" are deleted. Example 2.9 shows that this is false.

4.6. Two-Person Zero-Sum Games

A game is called *2-person 0-sum* if $n = 2$ and if there are utility functions u_1 and u_2 for the players such that

$$u_1(x) + u_2(x) = 0$$

for all $x \in X$. After adopting a specific such pair of utility functions, one defines the value of the game as usual. Specifically, Proposition 4.3 provides an equilibrium point in objective mixed strategies; by the minimax theorem, all such equilibrium points must have the same payoff, which is of the form $(v, -v)$. Then v is called the *value* of the game.

Proposition 6.1. *Let G be a 2-person 0-sum game with value v. Assume that*

there are outcomes x and y with $u_1(x) > v > u_1(y)$, and (6.2)

for each player i, there is a subjective event B_i regarding which i is informed. (6.3)

Then there is a pair $\boldsymbol{s}$ of strategies such that

$$H_1(\boldsymbol{s}) > v, H_2(\boldsymbol{s}) > -v. \tag{6.4}$$

Remark. The proposition is considerably easier to prove if one assumes that

there is a public subjective event B and there is a public roulette. (6.5)

To see this, assume without loss of generality that $p_1(B) > p_2(B)$ (otherwise, use the complement of B). If C is a public event and $0 \leqq \theta \leqq 1$, denote by θC a public event such that

$$p_1(\theta C) = \theta p_1(C) \quad \text{and} \quad p_2(\theta C) = \theta p_2(C);$$

the existence of such a θC follows[17] from Lemma 4.4. From (6.2) it follows that there is a β with $0 < \beta < 1$ such that

$$v = \beta u_1(x) + (1 - \beta)u_1(y).$$

For $0 < \delta < 1$, set

$$B^\delta = \Omega \backslash \delta(\Omega \backslash B).$$

Then for δ sufficiently small, we have $p_1(B^\delta) > p_2(B^\delta) > \beta$; hence if we let $A = \theta B^\delta$ for an appropriate θ, then

$$p_1(A) > \beta > p_2(A). \tag{6.6}$$

Now define $\boldsymbol{s}$ as follows: Both players choose pure strategies leading to x, or both players use pure strategies leading to y, according to whether A does or does not occur. Then (6.4) follows, and the proposition is proved under the assumption of (6.5).

[17]It also follows from the theorem of Lyapunov (1940) on the range of a vector measure. We shall see in Section 4.8 that Proposition 4.4 is itself proved via Lyapunov's theorem; intuitively, though, Proposition 4.4 is more transparent than Lyapunov's theorem.

Proof of Proposition 6.1. In (6.2), let $x = g(r_1, r_2)$ and $y = g(t_1, t_2)$; then $h_1(r_1, r_2) > v > h_1(t_1, t_2)$.

Suppose first that $h_1(r_1, t_2) \geqq v$. Setting

$$H^*(\varepsilon) = \varepsilon h_1(r_1, r_2) + (1 - \varepsilon) h_1(r_1, t_2)$$
$$H_*(\varepsilon) = \varepsilon h_1(t_1, r_2) + (1 - \varepsilon) h_1(t_1, t_2),$$

we find that for $\varepsilon > 0$ sufficiently small,

$$H^*(\varepsilon) > v > H_*(\varepsilon).$$

Hence there is a β in [0, 1] with

$$v = \beta H^*(\varepsilon) + (1 - \beta) H_*(\varepsilon).$$

Now let A_1 be a subjective event in $\mathscr{I}_1$ with $p_1(A_1) > \beta > p_2(A_1)$. The existence of such an event may be established as in the remark [cf. (6.6)], except that now the mapping $C \to \theta C$ takes $\mathscr{I}_1$ into itself (rather than $\mathscr{I}_1 \cap \mathscr{I}_2$ into itself), Lemma 4.2 (rather than 4.4) is used to prove the existence of θC, and B_1 is substituted for B. Again using Lemma 4.2, we obtain an objective 2-secret event A_2 with probability ε that is independent of A_1. Define a strategy pair $\boldsymbol{s}$ by stipulating that $\boldsymbol{s}_i$ takes the value r_i or t_i according to whether A_i does or does not occur. Then $\boldsymbol{s}$ is uncorrelated and satisfies (6.4).

If $h_1(r_1, t_2) \leqq v$, the proof is similar, the roles of the two players being then reversed. This completes the proof of Proposition 6.1.

Proposition 6.1 fails if it is only assumed that there is a subjective event of which at least one player is informed. For example, if in the 2-person 0-sum game with matrix

1	1
2	0

,

there is a subjective event in $\mathscr{I}_1$ but not in $\mathscr{I}_2$, then there is no pair of strategies satisfying (6.4). In this game it is sufficient for (6.4) that there be a subjective event in $\mathscr{I}_2$. In general, the proof of Proposition 6.1 shows that in any specific game, only one player need use a subjective strategy, while the other can use an objective one;

but which player it is that uses the subjective strategy may depend on the game.

It is perhaps worth noting that in Proposition 6.1, the strategies constituting $\boldsymbol{s}$ may be taken to be independent; indeed, the strategies constructed in the proof are independent.

4.7. Proof of the Propositions of Section 4.4

We start with a lemma that is basic to the proofs in this section.

Lemma 7.1. *Let $\mathscr{R}$ be a roulette, and let $B^1, \ldots, B^l$ be events. Then for any α between 0 and 1, there is an objective event in $\mathscr{R}$ with probability α that is independent of each of the B^k.*

Proof. Consider the $(1+l)n$-dimensional vector measure u on $(\Omega, \mathscr{R})$ defined by

$$\mu_i(A) = p_i(A), \quad i = 1, \ldots, n,$$

$$\mu_{kn+i}(A) = p_i(A \cap B^k), \quad i = 1, \ldots, n, \quad k = 1, \ldots, l.$$

That μ is non-atomic follows from the non-atomicity of the p_i on $\mathscr{R}$. By the theorem of Lyapunov (1940) on the range of a vector measure, the range of μ is convex. Now

$$\mu(\Omega) = (1, \ldots, 1, p_1(B^1), \ldots, p_n(B^1), \ldots, p_1(B^l), \ldots, p_n(B^l))$$

and

$$\mu(\phi) = (0, \ldots, 0).$$

Hence $\alpha\mu(\Omega)$ is in the range of μ, i.e. there is a set A in $\mathscr{R}$ such that $\mu(A) = \alpha\mu(\Omega)$. This means that $p^i(A) = \alpha$ for all i and

$$p_i(A \cap B^k) = \alpha p_i(B^k) = p_i(A)p_i(B^k)$$

for all i. The proof of the lemma is complete.

Proof of Lemma 4.1. By induction on the number m of members s_i of S_i for which $\sigma_i(s_i) > 0$. Let $\mathscr{R}_i$ be the i-secret roulette provided by Assumption II. As often in the case of inductive proofs, it is convenient to prove a somewhat stronger statement than is needed; here we show that

a strategy $\boldsymbol{s}_1$ can be chosen obeying Proposition 4.1; so that the events $\{\boldsymbol{s}_i = s_i\}$ are in $\mathscr{R}_1$. (7.2)

If $m = 1$ there is nothing to prove. Let $m > 1$, and suppose the proposition true for $m - 1$. Let $S_i = \{s_i^1, \ldots, s_i^m, \ldots\}$, where $\sigma_i(s_i^j) > 0$ if and only if $j \leqq m$. By the induction hypothesis, there are disjoint objective events $B^1, \ldots, B^{m-1}$ such that $p(B^j) = \sigma_i(s_i^j)/(1-\sigma_i(s_i^m))$. By Lemma 7.1 with $\mathscr{R} = \mathscr{R}_i$, there is an objective i-secret event B^m independent of each of $B^1, \ldots, B^{m-1}$, such that $p(B^m) = 1 - \sigma_i(s_i^m)$. Set $A^j = B^j \cap B^m$ for $j < m$, $A^m = \Omega \backslash B^m$, and $A^j = \phi$ for $j > m$; then $A^j \in \mathscr{R}_i$ for all j. Define $\boldsymbol{s}_i$ by $\boldsymbol{s}_i(\omega) = s_i^j$ if and only if $\omega \in A^j$. Then $\boldsymbol{s}_i$ satisfies (7.2). This completes the proof of Lemma 4.1.

Proof of Lemma 4.2. Use Assumption II, and apply Lemma 7.1 with $\mathscr{R} = \mathscr{R}_i$ and $l = 1$.

Before proving Proposition 4.3 we need a lemma.

Lemma 7.3. *Let $(\boldsymbol{s}_1, \ldots, \boldsymbol{s}_n)$ be an n-tuple of strategies, and let $s \in S$. For some $i \in N$, suppose that all the $\boldsymbol{s}_j$ except possibly $\boldsymbol{s}_i$ are mixed. Then*

$$\begin{aligned} p_i\{\boldsymbol{s} = s\} &= p_i\{\boldsymbol{s}_i = s_i\} p_i\{\boldsymbol{s}_j = s_j \ \textit{for all}\ j \neq i\} \\ &= p_i\{\boldsymbol{s} = s_1\} \ldots p_i\{\boldsymbol{s}_n = s_n\}. \end{aligned}$$

Proof. W.l.o.g. let $i = 1$, and let $A_j = \{\boldsymbol{s}_j = s_j\}$ for all j; then

$$\{\boldsymbol{s} = s\} = A_1 \cap \ldots \cap A_n.$$

Moreover, $A_1 \in \mathscr{I}_1$ and $A_j \in \mathscr{S}_j$ for $j > 1$. By definition of S_j, each A_j with $j > 1$ is 1-independent of $A_{j+1} \cap \cdots \cap A_n$. Hence

$$\begin{aligned} p_1(A_2 \cap \cdots \cap A_n) &= p_1(A_2) p_1(A_3 \cap \cdots \cap A_n) \\ &= \cdots = p_1(A_2) \cdots p_1(A_n). \end{aligned}$$

Again by definition of $\mathscr{S}_j$, each A_j with $j > 1$ is 1-independent of $A_1 \cap A_2 \cap \cdots A_{j-1}$. Hence by the previous equation,

$$p_1(A_1 \cap A_2 \cap \cdots \cap A_n) = p_1(A_1 \cap A_2 \cap \cdots \cap A_{n-1})p_1(A_n)$$
$$= \cdots = p_1(A_1)p_1(A_2)\cdots p_1(A_n) = p_1(A_1)p_1(A_2 \cap \cdots \cap A_n).$$

This completes the proof of the lemma.

Corollary 7.4. *Let* $(\boldsymbol{s}_1, \ldots, \boldsymbol{s}_n)$ *be an n-tuple of mixed strategies. Then the* $\boldsymbol{s}_i$ *are independent.*

Proof of Proposition 4.3. If $\boldsymbol{s} = (\boldsymbol{s}_1, \ldots, \boldsymbol{s}_n)$ is an objective mixed equilibrium point, then for each strategy $\boldsymbol{t}_i$ of i,

$$H_i(\boldsymbol{s}) \geqq H_i(\boldsymbol{s}_1, \ldots, \boldsymbol{s}_{i-1}, \boldsymbol{t}_i, \boldsymbol{s}_{i+1}, \ldots, \boldsymbol{s}_n). \tag{7.5}$$

In particular, this holds when $\boldsymbol{t}_i$ is itself objective and mixed. From this it follows that $H(\boldsymbol{s})$ is a Nash equilibrium payoff.

Conversely, if $F(\sigma)$ is a Nash equilibrium payoff, then from Lemma 4.1 and Corollary 8.4 it follows that there is an n-tuple $\boldsymbol{s} = (\boldsymbol{s}_1, \ldots, \boldsymbol{s}_n)$ of objective mixed strategies with payoff $F(\sigma)$ that is in equilibrium against deviations that are restricted to objective mixed strategies; i.e. that (7.5) holds when $\boldsymbol{t}_i$ is an objective mixed strategy. In particular (7.5) holds when $\boldsymbol{t}_i$ is "pure," i.e. takes on a particular value in S_i with probability 1. To show that $\boldsymbol{s}$ is an equilibrium in our sense, we must show that (7.5) holds when $\boldsymbol{t}_i$ is *any* strategy of i, i.e. that i cannot "correlate into" the strategies of the other players. Intuitively this follows from the fact that the $\boldsymbol{s}_j$ are mixed, i.e. pegged on j-secret events; formally, though, we must show that our definition of secrecy does indeed yield this result.

W.l.o.g. let $i = 1$. Write $S' = S_2 \times \cdots \times S_n$, and $\boldsymbol{s}' = (\boldsymbol{s}_2, \ldots, \boldsymbol{s}_n)$.

For any s_1 in S_1, write

$$H_1(s_1, \boldsymbol{s}') = E_1(h_1(s_1, \boldsymbol{s}')).$$

Since (7.5) holds when $\boldsymbol{t}_i$ is "pure," it follows that for all $s_1 \in S_1$,

$$H_1(s_1, \boldsymbol{s}') \leqq H_1(\boldsymbol{s}). \tag{7.6}$$

Then for an arbitrary strategy $\boldsymbol{t}_1$ of 1, it follows from Lemma 7.3 and (7.6) that

$$\begin{aligned} H_1(\boldsymbol{t}_1, \boldsymbol{s}') &= \sum_{s \in S} p_1\{(\boldsymbol{t}_1, \boldsymbol{s}') = s\} h(s) \\ &= \sum_{s_1 \in S_1} \sum_{s' \in S'} p_1\{\boldsymbol{t}_1 = s_1\} p_1\{\boldsymbol{s}' = s'\} h_1(s) \\ &= \sum_{s_1 \in S_1} p_1\{\boldsymbol{t}_1 = s_1\} \sum_{s' \in S'} p_1\{\boldsymbol{s}' = s'\} h_1(s) \\ &= \sum_{s_1 \in S_1} p_1\{\boldsymbol{t}_1 = s_1\} H_1(s_1, \boldsymbol{s}_2, \ldots, \boldsymbol{s}_n) \\ &\leqq \sum_{s_1 \in S_1} p_1\{\boldsymbol{t}_1 = s_1\}\, H_1(\boldsymbol{s}) = H_1(\boldsymbol{s}). \end{aligned}$$

This completes the proof of Proposition 4.3.

Proof of Lemma 4.4. Apply Lemma 7.1 with $l = 1$, taking $\mathscr{R}$ to be a public roulette.

Proof of Proposition 4.5. We first consider the equilibrium payoffs. Let $\boldsymbol{s}$ and $\boldsymbol{t}$ be equilibrium points, and let $0 \leqq \alpha \leqq 1$. Let $\mathscr{R}$ be the public roulette. By Lemma 7.1, there is an objective event A in $\mathscr{R}$ with probability α that is independent of all the events $\{\boldsymbol{s} = s\}$ and $\{\boldsymbol{t} = s\}$ for all s in S. Now let each player i play the strategy $\boldsymbol{r}_i$ defined as follows: if A occurs, play $\boldsymbol{s}_i$; if not, play $\boldsymbol{t}_i$. In other words, i plays a given pure s_i in S_i if and only if the event

$$[A \cap \{\boldsymbol{s}_i = s_i\}] \cup [(\Omega \backslash A) \cap \{\boldsymbol{t}_i = s_i\}] \tag{7.7}$$

occurs. This is indeed a strategy, since the event (7.7) is in the σ-field $\mathscr{I}_i$. It may then be verified that $\boldsymbol{r}$ is an equilibrium point and that

$$H(\boldsymbol{r}) = \alpha H(\boldsymbol{s}) + (1 - \alpha) H(\boldsymbol{t}),$$

so the set of equilibrium payoffs is convex. The proof that the set of feasible payoffs is convex is similar, so the proof of Proposition 4.5 is complete.

4.8. *A Posteriori* Equilibria

We would like to view an equilibrium point as a *self-enforcing agreement.* In the scenario described in Section 4.3, if the players agree on an equilibrium point $\boldsymbol{s}$ before stage (ii) (the stage at which information about ω is received), no player will want to renege [i.e. choose a pure strategy other than $\boldsymbol{s}_i(\omega)$] after stage (ii). More precisely, when making the agreement each player i assigns subjective probability 0 to the possibility that he will want to renege after receiving his information.

There is, however, a difficulty here. Though the subjective probability of wanting to renege is 0, this possibility is not entirely excluded; more important, it is quite possible that a player assigns positive probability to a *different* player's wanting to renege. In that case the equilibrium point can no longer be considered self-enforcing; the possibility of *somebody* wanting to renege is not negligible. This is exactly the phenomenon that is responsible for the equilibrium payoff of (1, 1) in Example 2.2.

To legitimize the view of an equilibrium point as a self-enforcing agreement, one can either

a. make assumptions under which the possibility of some player wanting to renege is assigned probability 0 by all players; or
b. construct a model in which it is possible to define equilibrium points at which no player *ever* wants to renege.

Specifically, we may assume that though the players may have different subjective probabilities, the concept of 'impossible' or 'negligible' is the same for all. That is, if $p_i(B) = 0$ for one i then $p_i(B) = 0$ for all i; in other words, the p_i are absolutely continuous with respect to each other. [For the case of two players, this is the same as (5.2).] In that case, the possibility that any player will want to renege is negligible for all players, and so can be safely ignored.

Alternatively, one could replace the concept of equilibrium point by that of '*a posteriori* equilibrium point' — a strategy n-tuple from which no player i *ever* wishes to move (unilaterally), even after receiving his information about nature's choice of ω. Formally, this can be done by adding the following to the 7 items that define a game (see Section 4.3).

(8) For each player i and each ω in Ω, a relation $\succsim_i^\omega$ on the space of lotteries on X (*the preference order of i given his information about ω*).

The relations $\succsim_i^\omega$ will also be called *a posteriori preferences*. An *a posteriori equilibrium point* is then defined to be an n-tuple $\boldsymbol{s}$ of strategies such that for all ω, all i, and all t_i in S_i, we have

$$g(\boldsymbol{s}) \succsim_i^\omega g(\boldsymbol{s}_1, \ldots, \boldsymbol{s}_{i-1}, \boldsymbol{s}_{i+1}, \ldots, \boldsymbol{s}_n).$$

Under appropriate assumptions, one can then prove the following results:

(8.1) Theorem 5.1 holds without the hypothesis of mutual absolute continuity (5.2), if "equilibrium point" is replaced by "*a posteriori* equilibrium point."

(8.2) If all the p_i are mutually absolutely continuous, then the set of payoffs to *a posteriori* equilibrium points coincides with the set of payoffs to equilibrium points.

The proof of (8.1) is like the proof of Proposition 5.1, except that a pure strategy s_{ij} is now said to "enter actively" into the strategy $\boldsymbol{s}_i$ if there is *some* ω for which $\boldsymbol{s}_i(\omega) = s_{ij}$, even if i assigns probability 0 to the set of all such ω. Note that from (8.1) it follows that (1, 1) is not an *a posteriori* equilibrium payoff in Example 2.2.

One can also prove *a posteriori* results that are analogous to Propositions 4.3 and 4.5.

A logically complete development of the *a posteriori* theory involves rewriting Assumptions I and II, redefining many of the concepts defined above (such as i-secrecy and mixed strategy), and making assumptions that relate preferences to *a posteriori* preferences.[18] The development therefore becomes somewhat lengthy. In view of result (8.2), it does not seem worthwhile, at least at this stage, to impose a formal description of the a posteriori theory on the reader.

4.9. Discussion

a. The fact that differing subjective probabilities can yield social benefit is perhaps obvious. Wagers on sporting events, stock market transactions, etc., though they can be explained by convexities in utility functions, can also be explained by differing subjective probabilities; and in reality, the latter explanation is probably at least as significant as the former. Correlation of strategies is basically also a fairly obvious idea. It is all the more surprising that these ideas have heretofore not been more carefully studied in game theory in general, and in the context of randomization in particular.

b. Our more substantive results can be broadly divided into two classes: Those having to do with equilibrium points (notably Examples 2.2 through 2.9, and Proposition 5.1), and those having to do with feasible points (notably Example 2.1 and Proposition 6.1). The former belong to the "non-cooperative" theory, the latter to the "cooperative" theory. An understanding of this distinction rests on an understanding of the concepts of communication, correlation, commitment, and contract. Let us examine these concepts in the context of this paper, noting in particular how they are related to "cooperative" and "non-cooperative" games.

---

[18]The *a posteriori* preferences of player i can be derived from his *a priori* preferences, but (like conditional probabilities) only for p_i-almost all ω. Of course, as we saw above, it is precisely the sets of p_i-measure 0 that cause the difficulty.

Recall that in step (ii) of the scenario of Section 4.3, each player i is given his information about nature's choice of ω. By *communication* we mean communication between the players before step (ii). Strategies of the players are *correlated* if they are statistically dependent. A *commitment* is an irrevocable undertaking on the part of a player, entered into before step (ii), to play in accordance with a certain strategy.[19] A *contract* (or "binding agreement" or "enforceable agreement") is a set of commitments simultaneously undertaken by several players, each player's undertaking being in consideration of those of the others.

Of the four terms just introduced, only "correlation" has a formal meaning in the framework of our formal model (see Section 4.3, where "uncorrelated" is defined). The other three refer not to the model itself, but to how the model is to be interpreted.

If the players can enter into commitments and contracts, we have a *cooperative game*. If not,[20] we have a *non-cooperative game*. Formally, cooperative and non-cooperative games are described in the same way, namely by the model of Section 4.3. The difference between the two types of games lies not in the formal description itself, but in what we have to say about it — what kind of theorems we prove about it.

In a non-cooperative game, at each point of time, each player acts so as to maximize his utility at that point of time, without taking previous commitments into account. Therefore the only strategy n-tuples of interest are those that are in equilibrium (cf. Section 4.8). On the other hand, in the cooperative theory the players can enter into binding agreements before step (ii); therefore a strategy n-tuple

---

[19]One can imagine broader meanings for the word 'commitment,' in which the undertaking, though irrevocable, need not be to play in accordance with a certain strategy. For example, it could be to play in accordance with one of a certain set of strategies, or it could be contingent on the choice of ω or on other commitments. We have not found it necessary to complicate the discussion by considering these broader meanings.

[20]These are the two extremes. In many situations, some commitments and contracts are possible, others not.

does not have to be in equilibrium to be of interest, and one is led to study all feasible n-tuples.

In Section 4.8 we saw that equilibrium points can be viewed as self-enforcing agreements. This is to be contrasted with feasible points that are not in equilibrium; if the players agree before step (ii) of the scenario to play such a point, some of them will wish to renege after step (ii). Therefore a contract — and an external enforcement mechanism — are required to make such agreements stick.

Thus we see that both the non-cooperative and the cooperative theory involve agreement among the players, the difference being only in that in one case the agreement is self-enforcing, whereas in the other case it must be externally enforced. Agreement usually involves communication, so that we conclude that communication normally takes place in non-cooperative as well as cooperative games.

As for correlation, this is sometimes taken as a hallmark of cooperative games. In our view, this is a fallacy; correlation may or may not be possible in a given game, whether or not it is cooperative. Examples 2.4 through 2.9 show that correlation may be involved in agreements that are not enforceable but are self-enforcing, so that it can be significant even in a non-cooperative game. On the other hand, even in a cooperative game it may be impossible to correlate, for example because the players are in different places and cannot observe the same random events; but nevertheless they can negotiate and even execute a contract, for example by mail.

Equilibrium points can be viewed in ways other than as self-enforcing agreements. For example, they can be viewed as strategy n-tuples with the property that if, for some extraneous reason, they "come to the fore" or are suggested to the players, there will be no tendency to move away from them. Alternatively, they can be viewed as providing only a necessary condition for a satisfactory theory of non-cooperative games: if a theory is to "recommend" a specific strategy to each player, then the n-tuple of the recommended strategies must be in equilibrium. To make sense, these points of view do not require pre-play communication. But even without communication, correlation is by no means ruled out. Thus in Example 2.4, the payoff (3/2, 3/2) is no more the result of communication than the payoff

(2, 1). All that is needed to achieve (3/2, 3/2) is that a fair coin be tossed, or a similar random device be actuated, and that the outcome be communicated to both players before the beginning of play. If that is done, the payoff (3/2, 3/2) becomes a full-fledged equilibrium payoff, conceptually indistinguishable from (2, 1); and that is so even in the absence of communication.

Of course, communication may be important in order to choose a particular equilibrium point from among all those whose payoff is (3/2, 3/2). But the problem of choosing a particular equilibrium point has nothing to do with correlation — it exists in most non-cooperative games. Thus in Example 2.4, even without the opportunity for correlation, it is difficult to see why one equilibrium point rather than another would be chosen. This, incidentally, is one reason for preferring to view an equilibrium point as a self-enforcing agreement.

Finally, a word about the distinction between "commitment" and "contract." Unlike a contract, a commitment is an *individual* undertaking; it may, for example, consist of giving irreversible instructions to an agent or a machine to act in accordance with a given randomized strategy, or it may involve an obligation to pay a large indemnity if one does not act in accordance with this strategy. If there is no such agent or machine, or nobody to enforce collection of the indemnity, it may be impossible to undertake commitments, and we will have a non-cooperative game. Commitments are important in game theory even when contracts are impossible [see for example Aumann and Maschler (1972)]. However, when it is possible to make commitments as well as to communicate, it should usually be possible to make contracts as well. Moreover, by permitting commitments one opens the door to pre-emptive tactics (such as threats) in the pre-play stage, which may easily lead the outcome away from equilibrium. Thus we feel that when one has admitted commitments, one has already gone much of the way from the non-cooperative to the cooperative theory.

To sum up: The cooperative theory requires communication as well as commitment and contracting power; and it is *a priori* concerned with all feasible outcomes. The non-cooperative theory requires that there be neither commitment nor contracting power, but it permits communication; and it is concerned with equilibrium outcomes. Both theories can accommodate correlation, but do not require its presence.

c. It is interesting that those of our results that belong to the "cooperative" theory are concerned with 2-person 0-sum games, which have traditionally been considered the epitome of non-cooperative games; it has been asserted that such a game can only be played non-cooperatively, since it can never be worthwhile to reach an agreement concerning it. Our results show that this is incorrect. It is not true that a 2-person 0-sum game is strictly competitive, i.e. that the preferences of the players are always in direct opposition;[21] both players can gain by the use of a binding agreement. Thus 2-person 0-sum games can profitably be played cooperatively.

On the other hand, when correlation is ruled out, the introduction of subjective strategies creates no new equilibrium payoffs [see (5.1)]. This indicates that subjectivity alone (without correlation) at least does not affect the non-cooperative theory of 2-person 0-sum games. However, we must be careful before reaching even this modest conclusion. The arguments for the use of minimax strategies, as presented by von Neumann and Morgenstern (1953), are of two kinds: the equilibrium arguments, which they (§17.3) call "indirect," and what they (§14, §17.4–§17.8) call "direct" arguments.[22] The indirect or equilibrium arguments are of course not affected by the introduction of subjectivity. But it is the direct arguments that have traditionally been used to support the contention that the theory of 2-person 0-sum games is conceptually more satisfactory than that of more general games. To a certain extent, these direct arguments *are* affected; implicitly, they depend on the strictly competitive character of the game, which as we know is removed by subjectivity.

---

[21]Nevertheless, even after the introduction of subjective strategies, the game remains "almost strictly competitive" in the sense of Aumann (1961).

[22]These are called "guaranteed value arguments" by Aumann and Maschler (1972).

d. It must be stressed that the chance elements of our model — the space $(\Omega, \mathscr{B})$ and the σ-fields $\mathscr{I}_i$ — are used for purposes of randomization only. Nature's choice of a point ω in Ω does not directly affect the outcome of the game; the outcome function g does not depend on ω. This model is therefore quite different from extensive game models in which chance makes "moves" that affect the outcome; here chance is only "made available" to the players for purposes of randomization. [See also subsection (g) below.]

e. Our use of the terms "subjective" and "objective" is a little different from the usual. The term "subjective" often signifies a personalistic definition of probability, based on preferences, such as that of Savage (1954), whereas "objective" signifies a physical (i.e. frequency) definition.[23] This is not the distinction being made here. Here the distinction is based solely on whether the players in the game do or do not agree about the numerical value of the probability in question.[24]

f. In this paper we have extended the theory of games in strategic (i.e. normal) form by the consideration of subjective events. A parallel extension is possible for the theory of games in extensive form. Basically, such an extension would consist of replacing the probabilities now appearing in the extensive form [Kuhn (1953)] on chance moves by *vectors* of probabilities, with one component for each player. The definition of "strategy" remains unchanged; the definition of "payoff"

[23]Cf. Anscombe and Aumann (1963), where personalistic and frequency probabilities were called "probability" and "chance" respectively.

[24]The fact that they may disagree forces us to adopt a personalistic definition of probability for at least some events; but there is nothing to prevent us from adopting a physical definition for other events (e.g. roulette spins) on whose probabilities the players agree. Hybrid systems in which the two kinds of probability exist side by side, have been investigated; cf. Anscombe and Aumann (1963). On the other hand, a purely personalistic view, such as Savage's, is also perfectly satisfactory for the purposes of this paper. Incidentally, if one does adopt such a purely personalistic view, then an *a priori* assumption on the existence of objective roulette wheels loses its intuitive attractiveness; thus the fact that Lemmas 4.1 and 4.2 follow from Assumption II gains in significance (see the discussion after the statement of Lemma 4.2).

is modified only in that one uses the different probabilities on the chance moves in calculating the payoffs for the different players. It may be verified that the basic theorems [Kuhn (1953)] of extensive game theory — namely the theorem on pure strategy equilibrium points in games of perfect information, and on behavior strategies in games of perfect recall — go through in this case as well.

One immediate application of this remark is to the theory of Harsanyi (1967, 1968) of games of incomplete information. Harsanyi has shown that in what he calls the "consistent case," a game of incomplete information in strategic form corresponds to a certain game of complete information in extensive form. In the "inconsistent case," there is no such correspondence. However, if one extends the definition of extensive games in the way indicated above, then the inconsistent case can be taken care of in essentially the same way as the consistent case.

g. An alternative approach to that of this paper would be to introduce the possibilities for correlation and subjective randomization explicitly into the extensive form of the game. This would mean starting the game with chance's choice of a point in Ω, and then giving each player i that information which is "his" in accordance[25] with the σ-field $\mathscr{I}_i$. The strategic form of this enlarged game could be calculated as indicated under (f) above; the Nash equilibrium points of the enlarged game correspond to the equilibrium points, as defined in this paper, of the original game. Thus in the appropriate context, our equilibrium points are special cases of Nash equilibrium points. One difficulty with this approach is that it would lead to an infinite extensive game, since Ω is infinite. But this is by no means an insuperable difficulty; models have been studied that are entirely adequate to cover such a situation [Aumann (1964)]. We did not construct our model in this way for a number of reasons. First, none of the examples or propositions would have been simplified by such a procedure; on the contrary, the extensive form is clumsy to

[25]The information sets of i immediately after chance's choice would be those subsets B of Ω such that for all A in $\mathscr{I}_i$, either $B \cap A = \varnothing$ or $B \subset A$.

work with, and the mathematical treatment would presumably have been more complex. Second, one usually thinks of the extensive form as representing the originally given rules of the game. In that case extraneous random events, which do not directly affect the outcome of the game, do not belong in the extensive form; one does not introduce all the possibilities for objective mixing explicitly into the game either [see subsection (e) above].

h. It is nevertheless useful to point out how some of our examples involving correlation look in an extensive framework. Example 2.4 may be restated as follows: Suppose we add a move at the beginning of the game in which chance simply announces the results of a coin toss; no other change is made in the game (see Figure 4.1). One would have thought that this could not possibly effect an essential change in the game. But in the classical theory of Nash (1951), it does; (3/2, 3/2) becomes the payoff to an equilibrium point, whereas it was not one before. Similarly if chance announces the result of a roulette spin, every point in the interval connecting (2, 1) to (1, 2) becomes an equilibrium payoff, whereas of these points, only the end points were equilibrium payoffs before. As we said in Section 4.2, all this appears already in Harsanyi and Selten (1972). But Examples 2.5 and 2.6 can be treated in a similar manner. Here one must add a move in which chance tosses a fair coin and informs players 1 and 2 — but not player 3 — of the outcome. Again, one would have thought that the addition of such a move could not possibly affect the game in any essential manner. But it adds a Nash equilibrium point with payoff (2, 2, 2), whereas in the original game no player could get more than 1 at any Nash equilibrium point. Examples 2.7 and 2.8 may be treated similarly.

i. The phenomena in Examples 2.1 and 2.3, which are the basic examples regarding subjective probabilities, do not depend on each player knowing the other players' subjective probabilities *precisely*. Thus in Example 2.1, it is sufficient that both players know that players 1 and 2 ascribe to D probabilities that are $>1/2$ and $<1/2$ respectively; in that case it will already be worthwhile to enter into a binding agreement. A similar remark holds for Example 2.3, in

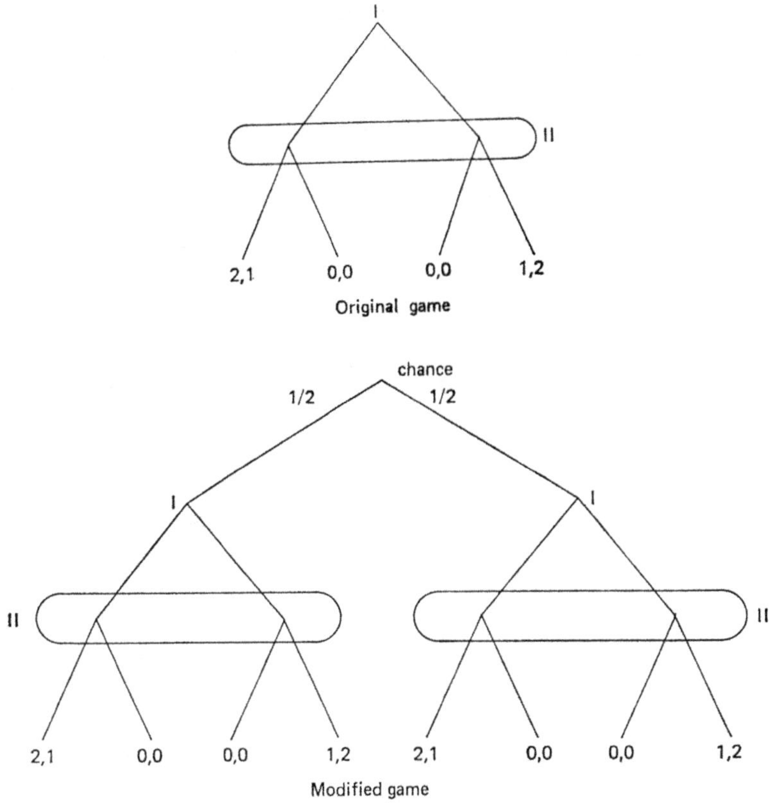

Figure 4.1

which it is sufficient that players 1 and 2 ascribe to D subjective probabilities that are *approximately* 3/4 and 1/4 respectively (the precise limits of the approximation are easily calculated).

Of course there is no particular reason to treat subjective probabilities differently from utilities — if we assume that the players' utilities are known to each other, we may as well assume the same for the subjective probabilities. The above remark only points out that precise knowledge is not crucial. Situations in which the players' utilities and/or subjective probabilities are not known to each other can be treated by the methods of games of incomplete information [Harsanyi (1967, 1968)].

j. The view is sometimes held that when people have different subjective probabilities for the same event, this can only be due to differences in the information available to these people. Such a view has been eloquently set forth by John Harsanyi (1968, §16), and we will call it the *Harsanyi doctrine.* Suppose, for example, that players 1 and 2 have subjective probabilities 2/3 and 1/3 respectively that a given horse A will run faster than another horse B in a given race. One could simply say that the subjective probabilities are different and leave it at that. But one could also imagine that both players previously had a uniform prior on the probability p that A beats B; that 1 had seen A beat B in one race, and 2 had seen B beat A in another race. Harsanyi's view is that differences in subjective probabilities can always be accounted for in such a fashion.[26]

The holders of such a view would probably consider the approach of this paper invalid. Suppose that in Example 2.1, D is the event "horse A beats horse B." Suppose further that each player knows that the other has observed exactly one previous race. Now in our context we assume that each player knows the other's probability for D. But in that case, the very knowledge of the other player's probability will cause revision of each player's own probability. The result of this revision will necessarily be $p_1(D) = p_2(D) = 1/2$, so that the previously "subjective" probabilities become "objective."

But there is also another possibility, namely that each player has made several observations, and that these observations lead the players to assign probabilities 2/3 and 1/3 respectively to D. For example, 2/3 would be the result of 3 wins for A and 1 win for B, as well as 1 win for A only. Though the players may know each other's probability, they may not know on precisely what observations it is based. Thus each player would not know if his own information is or is not more reliable than the other player's, and might be inclined to stick with his own information. Of course some revision of probabilities would certainly be called for even in such a case;

[26]Though Harsanyi (1968) does not state this position in such absolute terms, there is little doubt that that is his belief.

but whether such a revision must always ultimately lead to equal probabilities for D is not clear.

In any event, under the Harsanyi doctrine cases of differing subjective probabilities known to all players either do not occur, or if they do occur, they should be treated by the methods of games of incomplete information. Basically, therefore, that part of this paper dealing with subjective randomization assumes that different people can have irreconcilable priors — precisely what Harsanyi calls the "inconsistent case." Most workers in the field would probably agree that the inconsistent case can occur, and we are perfectly willing to let our contribution stand or fall on this basis.

But there is also another argument for our theory, an argument based on Savage's "Small Worlds" theme (1954, §2.5 and §5.5). Suppose we are faced with a game like that in Example 2.1 or 2.3, and wish to apply the Harsanyi doctrine; this leads to an analysis by means of the theory of games of incomplete information. Now such an analysis will in general involve an enormous expansion of the formal description of the game. It may be necessary to use a population of many millions of types for each of the 2 or 3 players.[27] The resulting game will in practice be completely unanalyzable. We suggest that an equally valid practice would be to accept apparent differences in subjective probabilities at their face value, even if one adheres to the Harsanyi doctrine.

This question is not too different from the question "What is an outcome?" (cf. Savage, op. cit.). Suppose we are playing a given matrix game for money. The normal procedure would be to assign utilities to the amounts of money involved, and then solve the game using any of the standard theories. But a sum of money is not in itself valuable; it depends on how one wishes to use it. So to analyze the game properly we would have to decide on how to invest the money or what consumer article to buy with it. But even the investment or the consumer article itself is often only a means to an end; and this

[27]... *every* possible combination of attributes ... will be represented in this population ... [Harsanyi (1968, p. 176), italics in the original].

"end" in turn, is often again largely a means. Following through all this in a formal fashion, even if it were theoretically possible, would make the simplest of games totally unmanageable. So one isolates the given game as a "small world," using the utilities to sum up all that follows. We suggest that even if in principle one accepts the Harsanyi doctrine, one can live with differing subjective probabilities as a summation of the complex informational situation in which the players find themselves.

k. We end this paper with a discussion of two possible objections that could be raised against the idea of subjective strategies. To fix ideas, consider the situation of two politicians, Adams and Brown, running for the office of mayor of their town. Each one has a number of (pure) campaign strategies open to him. Each pair of such strategies yields a probability for the election of Adams, the complementary probability being that of the election of Brown; for simplicity, assume that these probabilities are objective. This is a classical example of a constant-sum game, since there are only two possible final outcomes; the utilities of Adams and Brown may thus be taken equal to the respective success probabilities, and the sum is therefore always constant (=1).

What is now being suggested is that Adams and Brown use randomized campaign strategies that are pegged on events for which they have different subjective probabilities. These events may be entirely disconnected with the political campaign in question; to take an extreme example, if the candidates are Americans, they may peg their choices on the outcome of a cricket match in England.

One objection that may be raised to this procedure is that whereas a prudent person might be willing to use mixed strategies based on a coin toss with known objective probabilities, he would hesitate to risk his career on the outcome of an event of which he knows little or nothing. Adams and/or Brown may know little or nothing about cricket in England; how can we suggest that they peg their decisions on it?

Though this objection has great intuitive force, it is not consistent with the theory of subjective probability. According to this theory, people have subjective probabilities for any event, no matter how well or ill they are informed about it; of course the probability will depend on the available information, but there will always *be* a probability. Once this probability is determined, it enters the decision-making process in all respects like an objective probability. A certain amount of introspection will, moreover, support the conclusions of the theory despite their apparent strangeness. Suppose Adams is asked to choose between winning the election with objective probability 1/3 and winning the election if the Manchester cricket team beats the Liverpool team. If he knows nothing at all about the cricket situation, and subscribes to the "principle of insufficient reason," he might well choose the second possibility, despite — or rather because of — his lack of information. The same point may similarly be made by considering the situation in which it is not a pre-determined coin that is tossed, but rather a biassed coin that is picked out of a hat with a wide distribution of biasses; in such a situation, the utility theory of von Neumann and Morgenstern (1953) still applies, although in a certain sense the user knows much less about the coin.

1. Another possible objection runs as follows: Both players realize that the game is basically zero-sum, since in the last analysis, only one of them can win the election. Both know it is an illusion to think that both can gain more than the value of the game; this illusion is based on what might be called a mistaken appraisal of probabilities by at least one of them (and possibly both). Would they not be more prudent, then, to guarantee to themselves the objective value of this game by the use of objectively randomized strategies, thus avoiding all possible "mistakes"?

On one level, this question can be answered simply by saying that each side is here taking advantage of what it sees as the other side's mistake, and that this is perfectly rational. On a deeper level, though, it is incorrect to speak of "mistakes" at all. Each side has well-defined systems of preferences, each perfectly consistent in itself, and these preferences are *not* directly opposed. Thus, it is incorrect to say that the situation is "basically zero-sum." This situation is analogous to a situation in which the players' preferences as between *pure* outcomes are directly opposed, though between mixtures of outcomes they may not be; such a game is not usually called zero-sum or even "strictly competitive."

Consider, for example, the game

	L	R
T	5, 0	2, 2
B	1, 3	0, 5

Preference-wise, this matrix has a saddle point at (T, R). It is, however, not constant sum, since a contract to play (T, L) and (B, R) with probability 1/2 each is preferred by both players to the saddle point. Rather than signing the contract before tossing a coin, one could imagine that the coin has already been tossed, but that the players are not informed of the outcome. Paralleling the above argument, one could then say that both players realize that the game is "basically strictly competitive," since the preferences as between pure outcomes are directly opposed. Thus both know that it is an illusion to think that both can gain more than the 2 units assigned at the saddle point; this illusion is based on a mistaken appraisal of how the coin actually fell. Would they not be more prudent to guarantee a sure payoff of at least 2?

Certainly if they knew how the coin fell, there would be no point in using it to mix outcomes; but in the current situation, they both prefer the 1/2–1/2 combination of (T, L) and (B, R) to the saddle point (T, R). This agreement is therefore a perfectly natural one in spite of the fact that one of the players must necessarily lose out. Completely similar reasoning holds in the case of a game that is zero-sum in utilities, but that can be made non-strictly competitive by the appropriate use of subjectively randomized strategies.

Acknowledgement

Research supported in part by National Science Foundation grant GS-3269 at the Institute for Mathematical Studies in the Social Sciences, Stanford University, and in part by CORE, the Center for

Operations Research and Econometrics at the Catholic University of Louvain.

References

Anscombe, F. J. and R. J. Aumann, 1963, A definition of subjective probability, Ann. Math. Stat. 34, 210–212. [Chapter 2]

Aumann, R. J., 1961, Almost strictly competitive games, J. Soc. Ind. Appl. Math. 9, 544–550.

Aumann, R. J., 1964, Mixed and behavior strategies in infinite extensive games, in: M. Dresher, L. S. Shapley and A. W. Tucker, eds., Advances in game theory (Princeton University Press, Princeton) 627–650.

Aumann, R. J. and M. Maschler, 1972, Some thoughts on the minimax principle, Management Science 18, P-54–P-63.

Harsanyi, J., 1967–68, Games of incomplete information played by Bayesian players, Parts I–III, Management Science 14, 159–182, 320–334, 486–502.

Harsanyi, J. and R. Selten, 1972, A generalized Nash solution for two-person bargaining games with incomplete information, Management Science 18, P-80–P-106.

Kuhn, H. W., 1953, Extensive games and the problem of information, in: H. W. Kuhn and A. W. Tucker, eds., Contributions to the theory of games II (Princeton University Press, Princeton) 193–216.

Lyapunov, A., 1940, Sur les fonctions-vecteurs completement additives, Bull. Acad. Sci. USSR Ser. Math. 4, 465–478.

Nash, J. F., 1951, Non-cooperative games, Ann. of Math. 54, 286–295.

Savage, L. J., 1954, The foundations of statistics (John Wiley, New York).

Von Neumann, J. and O. Morgenstern, 1944, Theory of games and economic behavior (Princeton University Press, Princeton); 3rd ed., 1953.

Chapter 5

Values of Markets with a Continuum of Traders*,[1]

It is shown that in markets with a continuum of traders, the allocations associated with the Shapley value are the same as the competitive allocations.

5.1. Introduction

THE "EQUIVALENCE PRINCIPLE" for the core[2] states that the core of a perfectly competitive market coincides with the set of its competitive allocations. It is the object of this paper to establish a similar equivalence principle for the Shapley value [**5, 21**, and **23**], a game theoretic concept quite different from the core. Since the Shapley value can be interpreted in terms of "marginal worth", it is perhaps *the* game theoretic concept most closely related to traditional economic ideas.

---

*Reprinted from *Econometrica* (1975), 43(4), 611–646, with permission from Wiley.

[1]Text of the Walras-Bowley Lecture delivered at the winter meeting of the Econometric Society held in Toronto in December, 1972. This work was supported in part by National Science Foundation Grant GS-40104 at the Institute for Mathematical Studies in the Social Sciences, Stanford University, and in part by CORE, the Center for Operations Research and Econometrics at the Catholic University of Louvain.

[2]For a comprehensive treatment of the core equivalence principle, see Hildenbrand [**14**].

By a "perfectly competitive" market we mean one in which there are many traders, with each trader holding only a negligible proportion of the total resources of the economy. This situation may be modelled by a continuum of traders — specifically, by a non-atomic continuum, as in Aumann [**1**].

We will be dealing with ordinary Walrasian pure exchange economies; no assumption of transferable utility will be made. The value was originally defined [**21**] for games with transferable utility only, but was later [**23**] adapted to apply also to games without transferable utility. It is this adaptation, itself adapted to non-atomic markets, that we shall use in this paper. Values are ordinarily expressed in units of utility; since we are dealing with markets that are defined *a priori* in terms of preferences rather than in terms of utilities, we shall wish to refer directly to allocations rather than to utilities. Allocations corresponding to values in such a set-up will be called "value allocations"; a precise definition will be given in Section 5.5 below.

To establish the equivalence principle for the value, we will have to make certain differentiability assumptions on the traders' preferences, assumptions that are not necessary for the core equivalence principle. Preferences satisfying these assumptions will be called "uniformly smooth." This notion is closely related to Debreu's "smooth preferences" [**10**]. We might say that the preferences in a market are uniformly smooth if and only if they are smooth in Debreu's sense and the properties embodying this smoothness hold uniformly over all the traders in the market. This, too, will be spelled out precisely below (Sections 5.4 and 5.5).

Recall that an allocation $\boldsymbol{x}$ is called *competitive* if there is a price vector p such that $(p, \boldsymbol{x})$ is a competitive equilibrium. We are now ready for the statement of our main theorem:

Theorem 1 (Value Equivalence Theorem). *In a non-atomic market with uniformly smooth preferences, the set of value allocations coincides with the set of competitive allocations.*

It should be noted that this result depends strongly on the non-atomicity of the market, i.e., on its perfectly competitive nature. The theorem is not true *in either direction* for finite markets.[3] This is to be contrasted with the situation for the core: competitive allocations are in the core for any market, even a finite one; there the continuum is needed only to establish the converse, that all core allocations are competitive.

The Value Equivalence Theorem can be interpreted in terms of social welfare functions, and the concept of an individual's contribution to society. Given utility functions for the traders, we may, for any allocation $\boldsymbol{x}$ and coalition S, consider the sum of the utilities of the members of S for the bundles assigned to them under $\boldsymbol{x}$. Let us think of this sum as the "welfare" of the coalition S under the allocation $\boldsymbol{x}$; in other words, assume the existence of an additively separable social welfare function consistent with individual preferences. The "marginal contribution" of an individual trader is then the amount of welfare added to society by his presence — i.e., the difference between the maximum levels of welfare that society as a whole can achieve, with that trader and without him. It can be shown (see the Appendix) that, under the conditions of Theorem 1, *the marginal contribution of a trader is almost surely the same as his "true" contribution*: If the traders are ordered at random, then the marginal contribution of a trader almost surely equals the amount added by him to the maximum welfare achievable by the coalition of traders preceding him in the random order. This justifies our reference to the marginal contribution of a trader simply as his "contribution." Our theorem then makes two assertions: First, that an allocation is competitive if its utility to each trader is precisely his contribution; and conversely, that for each allocation $\boldsymbol{x}$, individual utility functions can be chosen (consistent with the preferences), so that the contribution of each trader is precisely the utility of the bundle assigned to him by $\boldsymbol{x}$.

In brief, *an allocation is competitive if and only if for some choice of the individual utility functions, it yields to each trader precisely his contribution to the welfare of society.*

[3] Counter-examples may be found toward the end of Section 5.11.

Among the precursors of this result are results of Shapley [**22**], Champsaur [**6**], and Aumann and Shapley [**5**, Theorem J and Proposition 31.7]. The literature will be more thoroughly discussed in Section 5.12.

The next two sections are devoted to a review of the Shapley value, first for finite games and then for games with a continuum of players. Section 5.4 is devoted to the relevant definitions of differentiability and smoothness. In Section 5.5 the market model is presented, "value allocation" is defined, and the main results are stated. In Section 5.6 we again discuss the intuitive content of the main theorem, in a more leisurely manner than above. Section 5.7 is devoted to an illustrative example, and Section 5.8 to a suggestive but unrigorous demonstration of the main results, based on the calculus of infinitesimals. In Section 5.9 we discuss the intuitive meaning of the differentiability and smoothness conditions, and Section 5.10 is devoted to a similar discussion of the uniformity conditions. In Section 5.11 we discuss the extraordinary relation, implicit in our results, between the basically cardinal concept of Shapley value and the ordinal concept of competitive equilibrium. Section 5.12 is devoted to a discussion of the literature, and Sections 5.13 through 5.15 to proofs. An appendix clarifies the precise relationship between the marginal and "true" contributions of a trader.

Readers who wish to skim through the paper should perhaps read only Sections 5.5 and 5.8, and the intuitive discussion immediately following the statement of the Value Equivalence Theorem in Section 5.1.

5.2. The Value in Finite Games

Let N be a finite set; call the members of N *players* and the subsets of N *coalitions*. A *game on* N is a function v that associates with each coalition S a real number $v(S)$ (the *worth* of S), such that $v(\varnothing) = 0$. A *payoff vector on* N is a measure on the subsets of N. Intuitively, a payoff vector is simply a function that assigns a real number (the payoff) to each player; such functions are in an obvious one-to-one correspondence with the measures on N, and for reasons that are both technical and conceptual,[4] it is convenient to treat a payoff vector as a measure.

A *null player* in a game v is a player i such that $v(S \cup \{i\}) = v(S)$ for all coalitions S. Players i and j are called *substitutes* if $v(S \cup \{i\}) = v(S \cup \{j\})$ whenever S contains neither i nor j. For a fixed player set N, a *value* is a function φ that associates with each game v a payoff vector φv satisfying the following conditions:

Condition 2.1 (Additivity). $\varphi(v + w) = \varphi v + \varphi w$.

Condition 2.2 (Symmetry). $(\varphi v)(\{i\}) = (\varphi v)(\{j\})$ *whenever* i *and* j *are substitutes.*

Condition 2.3 (Efficiency). $(\varphi v)(N) = v(N)$.

Condition 2.4 (Null player condition). $(\varphi v)(\{i\}) = 0$ *whenever* i *is a null player.*

Proposition 2.1. *For each finite player set* N *there is one and only one value* φ*; it is given by the formula* $(\varphi v)(\{i\}) = E(v(S_i \cup \{i\}) - v(S_i))$*, where* S_i *is the set of all players preceding* i *in a random order on* N*, and* E *is the expectation operator when all* $|N|!$ *such orders are assigned equal probability.*

For a proof of Proposition 2.1 that is even simpler than the original proof [**21**], see [**5**, Appendix A].

5.3. The Value in Continuous Games

In this section we take for the underlying player space a measurable space $(T, \mathscr{C})$, i.e., a set T together with a σ-field $\mathscr{C}$ of subsets of T. The members of $\mathscr{C}$ are *coalitions*, but one should not think of the points t of T as individual players; rather, one should think of an individual player as an infinitesimal subset dt of T (cf. [**5**, Section 29]).

A *game on* $(T, \mathscr{C})$ is a function v from $\mathscr{C}$ to the real numbers such that $v(\varnothing) = 0$. There are various ways of extending the definition of value from the finite games studied in Section 5.2 to the more general

[4]We wish to think of the payoff to a coalition, not only to an individual.

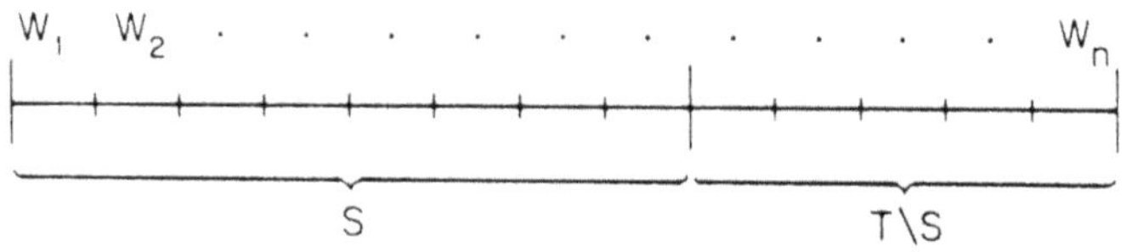

Figure 5.1

situation we have here; they are studied in detail in [**5**]. Here we will adopt a definition due to Kannai [**15**] (see also [**5**, Section 17]).

Let v be a game and S a coalition; we wish to define $(\varphi v)(S)$. Intuitively, one proceeds by dividing both S and $T\backslash S$ into a large but finite number of "small" sets W_i, say n in all (see Figure 5.1). If one considers each of these small sets as an individual player, then one gets a finite game by restricting v to unions of the W_i. In this finite game the coalition S — or more precisely the coalition of those W_i whose union is S — has a value, which we denote $\varphi_n(S)$. Now if we let $n \to \infty$ and the W_i shrink, $\varphi_n(S)$ may or may not tend to a limit. If it does, and if this limit is independent of the various choices that must be made, then we denote the limit $(\varphi v)(S)$, and call it the "asymptotic value of S."

Formally, let $\Pi_1, \Pi_2, \ldots$ be a sequence of partitions of T into measurable sets (i.e., members of $\mathscr{C}$), such that each Π_m refines the previous partition Π_{m-1}. Assume moreover that the sequence is *separating*, i.e., that if s and t are distinct points in T, then for m sufficiently large, s and t are in different members of Π_m; this is a precise expression of the idea that the sets in the partitions shrink. For example, if Π_m is the partition of the unit interval [0, 1] into the subintervals $[0, 1/2^m]$, $(1/2^m, 2/2^m], \ldots, (1 - 1/2^m, 1]$, then the sequence $\{\Pi_m\}$ is separating. For each m, let v_m be the finite game on Π_m defined by $v_m(\Xi) = v(\bigcup_{w\in\Xi} W)$, and let φv_m be its value. Let $\Pi_1 = \{S, T\backslash S\}$, and for each m, let $S_m = \{W \in \Pi_m\colon W \subset S\}$; S_m is the coalition in the finite game v_m corresponding to S in the original game v. Now let $m \to \infty$; if (φv_m) (S_m) has a limit, and if this limit is independent of the choice of the sequence $(\Pi_1, \Pi_2, \ldots)$ (when chosen in accordance with the above conditions), then the limit is denoted $(\varphi v)(S)$. If, moreover, this is the case of all $S \in \mathscr{C}$, then φv is called the *asymptotic value* of v. Note that the asymptotic value, if it exists, is a finitely additive measure on $(T, \mathscr{C})$. Note also that when T is finite, the asymptotic value coincides with the finite value.

The asymptotic value obeys conditions analogous to Conditions (2.1) through (2.4); see [**5**, Sections 2 and 17, in particular Theorem F]. Specifically, we note the *efficiency condition*:

$$(\varphi v)(T) = v(T). \tag{3.1}$$

5.4. Differentiability, Smoothness, and Uniform Smoothness

We will be working in a Euclidean space E^n. Superscripts will be used to denote coordinates. For x and y in E^n we write $x > y$ when $x^i > y^i$ for all i, and $x \geqq y$ when $x^i \geqq y^i$ for all i. The symbol $x \cdot y$ will denote the scalar product $\Sigma_i x^i y^i$. We will also use standard matrix notation, so that $x \cdot y = xy^*$, where both x and y are considered row vectors and * denotes "transpose." The symbol 0 denotes the origin of E^n as well as the number 0; no confusion will result. The non-negative orthant $\{x \in E^n \colon x \geqq 0\}$ in E^n is denoted Ω.

Unless otherwise specified, the word "function" will mean "real-valued function."

A function u on Ω is *continuously differentiable* (on Ω) if it can be extended to an open neighborhood (in E^n) of Ω on which it is continuously differentiable. An equivalent definition [**29**] is that u is continuous on Ω and continuously differentiable in the interior of Ω, and each of the partial derivatives of u can be continuously extended from the interior of Ω to Ω. In that case, the row vector of the partial derivatives of u is denoted u'; on the boundary of Ω, u' is defined to be the unique continuous extension of u' in the interior. A function on Ω with values in a finite dimensional Euclidean space is *continuously differentiable* (C^1 for short) if each of its components is continuously differentiable. A real-valued function u on Ω is *twice continuously differentiable* (C^2 for short) if both u and u' are continuously differentiable; the matrix of second derivatives is denoted u''.

A function u on Ω is called *quasi-concave* if for all $\alpha \in [0,1]$ and all x and y in Ω, $u(\alpha x + (1-\alpha)y) \geqq \min(u(x), u(y))$.

If u is a C^2 function on Ω, define

$$u^H = \begin{vmatrix} u'' & u'^* \\ u' & 0 \end{vmatrix}, \tag{4.1}$$

where * denotes "transpose." This is the *bordered Hessian* of u, i.e., the $(n+1) \times (n+1)$ determinant in which the $n \times n$ matrix u'' is bordered by the row vector u', by the column vector u'^*, and by the number 0. When $u' > 0$, define

$$u^c = \frac{-u^H}{\|u'\|_2^{n+1}}, \tag{4.2}$$

where $\| \ \|_2$ is the Euclidean norm. It may be verified [**10**] that u^c depends only on the level (or indifference) surfaces of u, and not on u itself. The *Gaussian curvature* of the level surface of u at x is then defined as $u^c(x)$.

When $n = 2$, then $u^c(x) = 1/r$, where r is the radius of the circle that "best fits" the level curve of u at x. Thus if $u^c(x) = 0$, then $r = \infty$, so that the "circle" that best fits the level curve is a straight line. Further discussion of the Gaussian curvature can be found in Section 5.9 and also in [**10**].

Let $(T, \mathscr{C})$ be a measurable space. For each $t \in T$, let w_t be a function from Ω to some Euclidean space E^l (possibly $l = 1$). The w_t are *uniformly bounded* if there is a point b in E^l such that $w_t(x) < b$ for all t in T and x in Ω. They are *uniformly bounded in compact sets* if for each compact subset C of Ω, there is a point b in E^l such that $w_t(x) < b$ for all t in T and x in C. They are *uniformly positive* (*or bounded away from zero*) *in compact sets* if for each compact subset C of Ω, there is a point $b > 0$ in E^l such that $w_t(x) > b$ for all t in T and x in C. They are *simultaneously measurable* if $w_t(x)$ is measurable as a function of (t, x), i.e., measurable in the product field $\mathscr{C} \times \mathscr{B}$, where $\mathscr{B}$ is the σ-field of Borel sets in Ω.

A family $\{u_t\}$ of functions on Ω is called *bounded differentiable* if the u_t are simultaneously measurable, uniformly bounded, and continuously differentiable on Ω, and the derivatives u'_t are, in compact sets, uniformly bounded and uniformly positive. Finally, $\{u_t\}$ will be called *uniformly smooth* if it is bounded differentiable and the following conditions are satisfied:

Condition 4.1. *The u_t are quasi-concave and twice continuously differentiable on Ω.*

Condition 4.2. *The Gaussian curvatures u_t^c are uniformly positive in compact sets.*

Condition 4.3. *The second derivatives u_t'' are uniformly bounded in compact sets.*

5.5. Description of Model and Statement of Results

The market model is basically the same as in [**1**]. It consists of: (i) a measurable space $(T, \mathscr{C})$, called the *trader space*, together with a σ-additive, non-negative measure μ on $\mathscr{C}$ with $\mu(T) < \infty$; (ii) the non-negative orthant Ω — called the consumption set — of a Euclidean space E^n, whose dimension n represents the number of different goods being traded in the market; (iii) a measurable function $\boldsymbol{a}$ from T to Ω called the *initial allocation*; and (iv) for each t in T, a relation $\succ_t$ on Ω, called the *preference relation* of t. The system consisting of these four items is called a *market*. The market is called *non-atomic* if μ is non-atomic.

We will assume that the measurable space $(T, \mathscr{C})$ is *standard*, i.e., that it is finite, denumerable, or isomorphic[5] to the unit interval [0,1] with the Borel sets. This assumption is less restrictive than it sounds; any Borel subset of any Euclidean space (or indeed, of any complete separable metric space) has this property.

If $\boldsymbol{f}$ is a function on T, we often write $\int_S \boldsymbol{f}$ instead of $\int_S \boldsymbol{f}(t)\mu(dt)$, and $\int \boldsymbol{f}$ instead of $\int_T \boldsymbol{f}$. We will make the following assumptions:

Assumption 5.1. $\int \boldsymbol{a} > 0$.

Assumption 5.2. *$\boldsymbol{a}$ is uniformly bounded.*[6]

[5]Two measurable spaces are *isomorphic* if there is a one-to-one transformation from one onto the other that preserves measurability in both directions.
[6]That is, there is a point b in Ω such that $\boldsymbol{a}(t)' < b$ for all t.

Assumption 5.1 is fairly standard; it means that each commodity is actually present in the market, though not all traders need initially hold a positive amount of it. Assumption 5.2 is less familiar; actually we make it only as a kind of normalization, and at the expense of certain complications below, could dispense with it. This matter will be discussed further in Section 5.10.

In the non-atomic case, it is best to view an individual trader as an "infinitesimal subset" dt of T, rather than as a single point in T. Individual bundles are also infinitesimal; thus the initial bundles of the individual trader dt is $\boldsymbol{a}(t)\mu(dt)$, whereas the point $\boldsymbol{a}(t)$ itself is to be thought of as dt's initial bundle *density* (initial bundle of dt divided by his measure). Similarly, $x \succ_t y$ means that dt prefers $x\mu(dt)$ to $y\mu(dt)$, i.e., he prefers the bundle density x to the bundle density y. This point of view is expounded at some length in [**5**, Section 29]; see also [**26**, Footnote 12].

A *utility* for a relation $\succ$ on Ω is a real-valued function u on Ω such that $u(x) > u(y)$ if and only if $x \succ y$. The preferences $\succ_t$ are called *uniformly smooth* if there exists a uniformly smooth family of utilities for them. An intuitive discussion of uniform smoothness will be given in Section 5.10.

Let $U = \{u_t\}$ be a bounded differentiable[7] family of utilities for the preferences. Define a game v_U on $(T, \mathscr{C})$ by

$$v_U(S) = \max\left\{\int_S u_t(\boldsymbol{x}(t))\mu(dt) : \int_S \boldsymbol{x} \right.$$
$$\left. = \int_S \boldsymbol{a} \text{ and } \boldsymbol{x}(t) \in \Omega \text{ for all } t \text{ in } S. \right. \qquad (5.1)$$

Intuitively, $v_U(S)$ is the maximum total utility that S can guarantee to itself by redistributing its total initial bundle $\int_S \boldsymbol{a}$ among its members. We remark that the maximum in the definition of v_U is attained [**4**], and the game v_U has an asymptotic value φv_U [**5**, Proposition 31.7].

[7] See Section 5.4.

An *allocation* is a measurable function $\boldsymbol{x}$ from T to Ω such that $\int \boldsymbol{x} = \int \boldsymbol{a}$. It is called a *value allocation* if there exists a bounded differentiable family U of utilities for the preferences such that for all t,

$$(\varphi v_U)(dt) = u_t(\boldsymbol{x}(t))\mu(dt), \tag{5.2}$$

i.e., such that for all S,

$$(\varphi v_U)(S) = \int_S u_t(\boldsymbol{x}(t))\mu(dt). \tag{5.3}$$

In words, (5.2) says that $\boldsymbol{x}$ yields each trader precisely the utility that he is assigned by the value. An allocation $\boldsymbol{x}$ is called *competitive* if there exists a p in Ω with $p \neq 0$ (called the vector of *prices*) such that for almost all t in T, $\boldsymbol{x}(t)$ is maximal with respect to $\succ_t$ in the *budget set* $\boldsymbol{B}_p(t) = \{x \in \Omega\colon p \cdot x \leqq p \cdot \boldsymbol{a}(t)\}$.

This completes the presentation of the basic definitions. The first result is:

Proposition 5.1. *In a non-atomic market, every value allocation is competitive.*

Conversely, we have:

Proposition 5.2. *In a non-atomic market, assume that either (i) the preferences are uniformly smooth, or (ii) there exists a bounded differentiable family of concave utilities for the preferences. Then every competitive allocation is a value allocation.*

Proposition 5.2 requires either smoothness or concavity; for Proposition 5.1 neither requirement is needed. But Proposition 5.1 does require differentiability, which is implicit in the definition of "value allocation".

The Value Equivalence Theorem — stated in the introduction — is, of course, an immediate consequence of Propositions 5.1 and 5.2.

We mention also that if in the definition of value allocation we replace "asymptotic value" by "mixing value" [**5**, Section 14, p. 115], then the Value Equivalence Theorem and Propositions 5.1 and 5.2 remain true without any change. The proofs are as in Sections 5.13

through 5.15; one merely has to replace the word "asymptotic" by the word "mixing".

5.6. Interpretation

The simplest interpretation is essentially the one sketched in the introduction (after the statement of the Value Equivalence Theorem). It involves an additively separable social welfare function, defined by a family $U = \{u_t\}$ of utilities. It is best to think of trader dt's utility for the bundle $x\mu(dt)$ as $u_t(x)\mu(dt)$. Then $\int_T u_t(\boldsymbol{x}(t))\mu(dt)$ is society's level of welfare under the allocation $\boldsymbol{x}$; and $v_U(S)$ is the maximum level of welfare that the coalition S can achieve by its own efforts. Thus $v_U(S)$ measures the power or bargaining strength of the coalition S; and *given the bargaining strengths*, φv_U may be considered the "equitable" payoff distribution. Hence a value allocation (see (5.2)) is one that yields each trader precisely his "equity" in the market; i.e., an allocation $\boldsymbol{x}$ whose utility $u_t(\boldsymbol{x}(t))\mu(dt)$ to each trader dt is precisely $(\varphi v_U)(dt)$.

Here we used the word "equity" in describing the value, whereas in the introduction we referred to the traders' "contributions". The assertion that the value constitutes an equitable payoff distribution is based on its axiomatic definition (see Conditions 2.1 through 2.4). That it also may be thought of in terms of the players' contributions is a consequence of Proposition 2.1 (see also Step 5 of Section 5.7, and the Appendix). These two interpretations are simply two sides of the same coin — two ways of looking at the same concept.

It should be noted that for a given family U of utilities, there will in general[8] be at most one such $\boldsymbol{x}$, and usually none; to obtain the set of all value allocations, one must allow U to vary. An illustrative example will be analyzed in Section 5.7.

There is a certain conceptual analogy between the process of determining a value allocation and that of determining a competitive allocation. To find a competitive allocation, one assigns prices to

[8]For example, whenever the u_t are strictly concave.

the commodities. Given the initial allocation, these prices generate individual demands. If the demands thus generated are feasible — i.e., if they can be obtained by reallocating the total initial supply — then we have a competitive allocation.

To find a value allocation, one assigns a utility function to each trader. Given the initial allocation, these utility functions generate a payoff distribution — the value. If this payoff distribution is feasible — i.e., if it can be obtained by reallocating the total initial supply — then we have a value allocation. The utility functions here play a role similar to that of the prices in the case of competitive allocations. Prices enable us to evaluate bundles of goods; utility functions enable us to evaluate coalitions of players. Workers in general equilibrium theory have in the past been loath to compare utilities; but there is some reason to believe that such comparisons play a significant role in economic activity. Workers in areas such as growth, income distribution, and taxation have recognized this and have been quite willing to assume social welfare functions and the utility comparisons they imply. The fact that there are certain conceptual difficulties involved in such comparisons does not permit us, ostrich-like, to ignore them. The analysis outlined here suggests that society may be generating utility comparisons by means of a mechanism similar to the familiar supply and demand mechanism that generates prices; more bluntly, that "social welfare" is a function of power rather than an ethical construct.

The above interpretation of the Value Equivalence Theorem — i.e., in terms of social welfare functions — is open to a certain objection; though not crippling, it bears discussion. The games v discussed in Sections 5.2 and 5.3 are usually interpreted as "side payment" or "transferable utility" (TU) games. Essentially, this means that the payoffs are assumed to be in a single, desirable commodity; that $v(S)$ is the *total* payoff that S can get for its members; and that S can get for its members *any* payoff totalling $v(S)$, as is for example the case when the players can freely transfer the single good among each other. The markets defined in Section 5.6, on the other hand, cannot be considered TU games: the payoff is in terms of the personal utilities u_t, which are not "commodities", and are certainly not transferable. How, then, can TU theory be relevant to these markets?

Naturally, one can answer that though the TU view of a game v is the usual one, it is not the only possible one, and that in the current application the social welfare view is preferred. But it is also possible to justify the notion of value allocation directly in terms of the TU view. One can think of markets in terms of production, as follows: there are n raw materials, not in themselves consumable, and one desirable consumer commodity, which for simplicity we shall call *food.* Each trader dt starts out with a bundle $\boldsymbol{a}(t)\mu(dt)$ of raw materials, and with no food. The traders may exchange raw materials. If, after the exchange, dt holds a bundle $x\mu(dt)$ of raw materials, he can produce from it an amount $u_t(x)\mu(dt)$ of food. Transfers of food are forbidden.

A little reflection will convince the reader that this is conceptually entirely equivalent to the usual view of a market.

Since food transfers are forbidden, this cannot be viewed as a TU game. But suppose now that food transfers are legalized.[9] Then each coalition S can produce an amount $v_U(S)$ of food, which it can divide among its members in any way it pleases. Therefore the situation is well represented by the TU game v_U.

In general, the value φv_U of v_U is not achievable without actual transfers of food (cf. Section 5.7). Specifically, we will have

$$(\varphi v_U)(dt) = u_t(\boldsymbol{x}(t))\mu(dt) + \zeta(dt), \tag{6.1}$$

where $\boldsymbol{x}$ is an allocation of resources that achieves $v_U(T)$, i.e., such that $v_U(T) = \int_T u_t(\boldsymbol{x}(t))\mu(dt)$, and $\zeta(dt)$ is the net intake of food by dt resulting from food transfers (rather than from his own production). These transfers clearly sum to 0, i.e., $\zeta(T) = \int_T \zeta(dt) = 0$; but for individual traders they will usually not vanish, i.e., the measure ζ will not vanish identically.

Suppose now that ζ does happen to vanish identically; then (6.1) reduces to (5.2), i.e., $\boldsymbol{x}$ is a value allocation. This means that the

[9]Economies of this kind (i.e., with legalized food transfers) are called *TU markets.*

value φv_U is achievable directly by means of the resource allocation $\boldsymbol{x}$, without any transfers of food. In this case $\boldsymbol{x}$ is called a *value allocation.*

If $\boldsymbol{x}$ is a value allocation, one can think of the traders going to market under the impression that food transfers will be permitted. This leads them to trade so as to arrive at the value of the associated game v_U; i.e., they will reach the allocation $\boldsymbol{x}$, but will undertake no subsequent food transfers. After going home, they hear that food transfers have been forbidden. Fortunately, nobody wanted to transfer food anyway. Thus, to arrive at a value allocation, one must only think about food transfers — imagine them — they don't actually have to be possible. The "transferable utility" does not really have to be transferable, because in the end nobody wants to transfer it.

Basically, this argument uses what is sometimes called the principle of independence of irrelevant alternatives (cf. [**18**]). We agree that when food may be transferred, then among all possible outcomes, an outcome that is in some sense best is to give to player dt the allocation $\boldsymbol{x}(t)\mu(dt)$. But this allocation does not require the transfer of food, and so *a fortiori* it should be best when such transfers are forbidden.

We close this section with a few words about the notion of "equity" and its connection with the Shapley value. We stated above that the value of a game v is calculated by taking into account the bargaining strengths of players and coalitions, as reflected in the worths $v(S)$. Thus the more a player can get for himself, acting alone or in concert with other players, the more the value will assign to him. Some readers may feel that this smacks more of power politics — of "might makes right" — than of what is commonly considered equity or fairness. Wouldn't it be more appropriate to adopt a point of view similar to that of Rawls [**20**] or of some of the recent literature on income redistribution,[10] which ignore the power structure of society and treat people in a basically egalitarian way?

We think not. Criteria of equity that do not take power considerations into account have a pleasing air of symmetry and

[10]See, for example, Sheshinski [**25**] and the papers quoted there.

abstract perfection; they appeal to our philosophical, ethical senses. But what would make society adopt such criteria? And even if adopted, what chance do they have of surviving the attacks of pressure groups, large and small? For better or for worse, economic realities are dictated not by ethical considerations, but by the exercise of economic power. We are not suggesting that the more egalitarian considerations are not worthy of study; but we do think that the kind of criterion exemplified by the Shapley value — which some readers might prefer to call a "reasonable compromise" rather than an "equitable solution" — is more relevant to economic reality.

5.7. An Example

This section is devoted to a trivial but nonetheless instructive example. Consider an economy in which there is only one commodity, which is desired by all traders (i.e., for all t, $x \succ_t y$ if and only if $x > y$). Let the initial bundle density be 1 for half the traders (say the set S_1), and 3 for the other half (say $S_3 = T \backslash S_1$). Of course one would not expect any trading to take place in such a situation, but let us follow through our formal analysis and look for value allocations. Let u be any strictly concave function on the nonnegative reals that is bounded, differentiable, and strictly increasing (e.g., $1 - 1/(x+1)$). Set $u_t = u$ for all t; for simplicity, let $\mu(T) = 1$. Then $v_U(S_1) = (1/2)u(1), v_U(S_3) = (1/2)u(3)$, and $v_U(T) = u(2)$. Since u is strictly concave, it follows that $v_U(T) > v_U(S_1) + v_U(S_3)$; in other words, trading is socially beneficial. To obtain the full benefit from trading, society must choose an allocation that actually yields a total utility of $v_U(T) = u(2)$. The only such allocation is $\boldsymbol{x}_0(t) \equiv 2$. But this cannot be a value allocation, as it is highly inequitable: it does not differentiate at all between the players in S_1 and S_3, and in fact assigns to traders in S_3 bundles smaller than their initial bundles! Hence there is no value allocation at all associated with this choice of the u_t — no allocation that is both socially optimal, and divides the benefits from trading equitably among the traders.

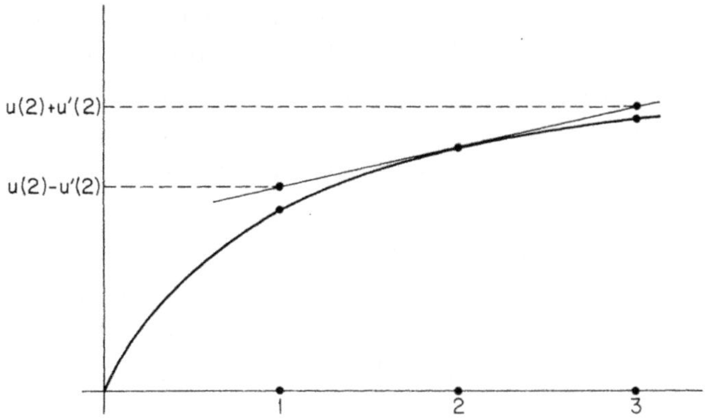

Figure 5.2

If one calculates[11] φv_U, one finds that it assigns densities $u(2) - u'(2)$ and $u(2) + u'(2)$ to the players in S_1 and S_3, respectively[12] (see Figure 5.2). But to achieve such a payoff distribution, one would have to start by assigning a density of 2 units of the single commodity to all traders, and then one would have to "transfer utility" in the amount of $u'(2)$ from S_3 to S_1 — an impossible (indeed meaningless) proceeding.

Since this choice of U does not lead to a value allocation, the question arises, which choice does? A trivial answer is provided by taking $u_t = w$ for all t, where $w(x) = x$ in an interval including $[1, 3]$. In that case, there is no social benefit from trading, and so the initial allocation is also a (in fact, *the)* value allocation.

But there is also a more instructive answer. Consider the u_t in our first example, i.e., $u_t(x) = 1 - 1/(x + 1)$. If we multiply these u_t by some $\alpha > 1$ for $t \in S_3$, and leave them unchanged for $t \in S_1$ then we are assigning a higher social benefit to consumption by S_3 than

[11] Step 5 of Section 5.8 or Lemma 13.2.

[12] When $u(x) = 1 - 1/(x + 1)$, then $u(2) = u'(2) = 5/9 = u(1) + 1/18$, and $u(2) + u'(2) = 7/9 = u(3) + 1/36$. Thus the value dictates a gain in utility (over that of the initial bundle) for all traders, but a smaller gain for S_3 than for S_1. "Equitable" is not necessarily "equal".

by S_1 — consumption by S_3 is considered "more valuable".[13] As a result, maximum social welfare will be attained by assigning a higher commodity density to S_3 than to S_1. In particular, if α is sufficiently high, maximum social welfare will be attained by transferring goods from S_1 to S_3. We have already seen that if $\alpha = 1$, then maximum social welfare is attained by transferring from S_3 to S_1. For an appropriate intermediate α, therefore, maximum social welfare is attained precisely at the initial allocation, i.e., without any transfers. Of course this does not necessarily imply that the value is then also achieved without any transfers, but it may be verified that that is in fact the case.

When there is more than one commodity, value allocations will in general involve transfers of commodities; but the utilities must be chosen so that no "transfer of utility" is required.

5.8. An Unrigorous Demonstration of the Results

The proof of our results relies heavily on the theory of non-atomic games developed in Aumann–Shapley [**5**]; when spelled out, it is long and involved. On the other hand, it is comparatively easy to "demonstrate" the theorem in terms of the calculus of infinitesimals and one or two other rough conceptual tools; that is what we shall do in this section. Such a demonstration, though unrigorous, gives insight into the situation and shows why one would expect the theorem to hold.

In Steps 1 through 5 of the demonstration, $U = \{u_t\}$ is a fixed bounded differentiable family of utilities, and $\boldsymbol{x}$ is an allocation at which $v_U(T)$ is achieved, i.e., such that

$$v_U(T) = \int_T u_t(\boldsymbol{x}(t))\mu(dt). \tag{8.1}$$

---

[13]Because the members of S_3 are more powerful, in that their initial bundles are greater.

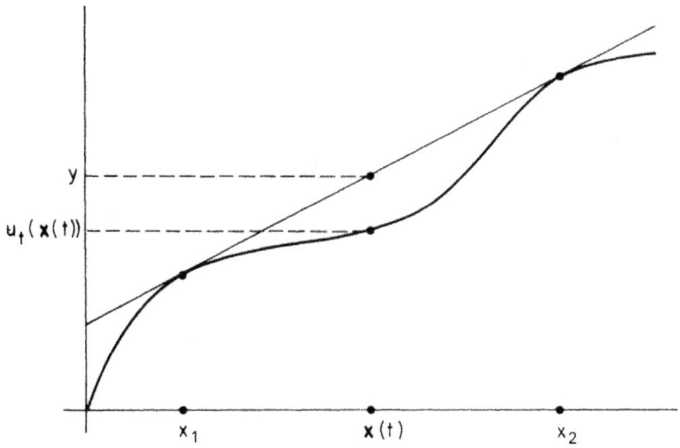

Figure 5.3

The universal quantifier ("for all t") is to be understood in each one of these steps.

Step 1: $(\boldsymbol{x}(t),\ u_t(\boldsymbol{x}(t)))$ *is on the boundary of the convex hull of the graph of* u_t.

When $n = 1$, this asserts that the situation cannot be as pictured in Figure 5.3. Indeed, if it were, we could split dt into two players, dt_1 and dt_2, and give them densities x_1 and x_2, respectively. The total utility that dt would get in this way would be $y\mu(dt)$ rather than $u_t(\boldsymbol{x}(t))\mu(dt)$. Thus dt, and so also T as a whole, would benefit from the split; this contradicts the efficiency of the value. The argument for arbitrary n is similar, except that it may be necessary to split dt into n traders $dt_1, \ldots, dt_n$. We are of course making strong use of the non-atomicity of the market in this step.

Step 2: $u_t'(\boldsymbol{x}(t)) = u_s'(\boldsymbol{x}(s))$ *for all* s *and* t.

Since the demonstration is in any case unrigorous, we assume without twinges of conscience that we are in the interior,[14] i.e., that

[14]Of course no such assumption is used in the rigorous proof. Even the statement of the step is false without the "interior" assumption, and must be replaced by a more complicated statement.

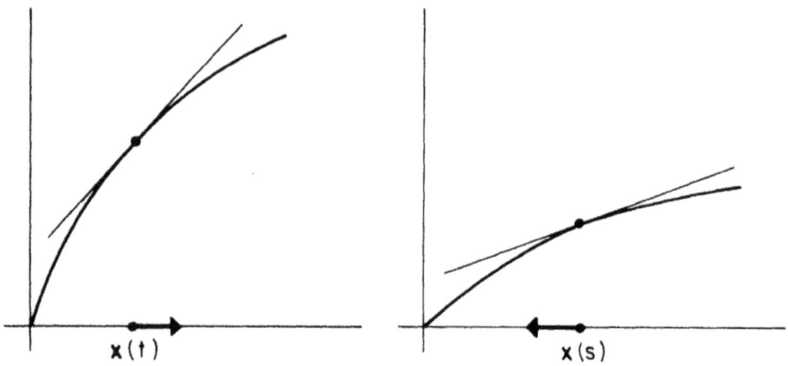

Figure 5.4

$\boldsymbol{x}(t) > 0, \boldsymbol{x}(s) > 0$. Let $n = 1$. If the assertion were false, say $u_t'(\boldsymbol{x}(t)) > u_s'(\boldsymbol{x}(s))$, then society could gain by moving $x(t)$ to the right and $x(s)$ to the left, in contradiction to the efficiency of the value (see Figure 5.4). The argument is similar for general n.

Set $p = u_t'(\boldsymbol{x}(t))$; by Step 3, p does not depend on t.

Step 3: *For all* x *in* $\Omega, u_t(\boldsymbol{x}(t)) - u_t(x) \geqq p \cdot (\boldsymbol{x}(t) - x)$.

From Step 2 it follows that the hyperplane with slope p through the graph of u_t at $x = \boldsymbol{x}(t)$ is tangent to the graph there; from Step 1 that this hyperplane supports the graph. This is precisely what is asserted here (see Figure 5.5).

Step 4: $\boldsymbol{x}(t)$ *is maximal with respect to the preference relation* $\succ_t$ *in* $\{x\colon p \cdot x \leqq p \cdot \boldsymbol{x}(t)\}$.

Indeed, if $p \cdot x \leqq p \cdot \boldsymbol{x}(t)$ then from Step 3 it follows that $u_t(x) \leqq u_t(\boldsymbol{x}(t))$.

Step 4 asserts that the prices p are *efficiency prices.* When the u_t are concave, Steps 2, 3, and 4 are well-known even in the finite case; the p^i are the "shadow prices" in the maximization that defines $v_U(T)$. The non-atomicity enables the establishment of Step 1, and so the extension to non-concave u_t.

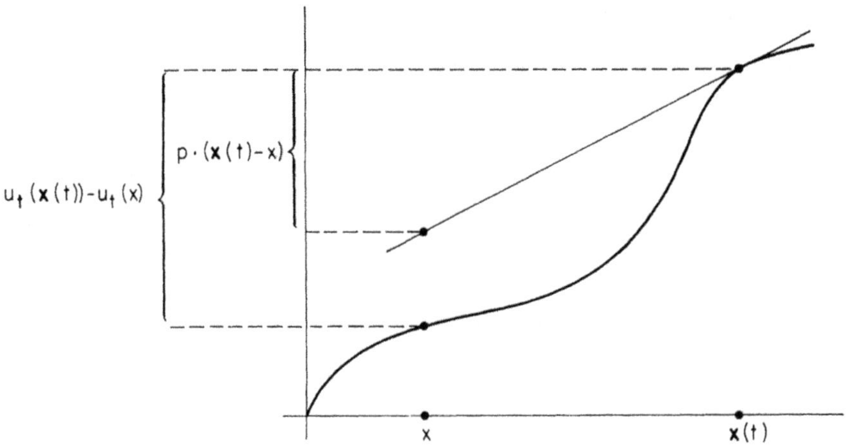

Figure 5.5

The following step is crucial:

Step 5: $(\varphi v_U)(dt) = (u_t(\boldsymbol{x}(t)) + p \cdot (\boldsymbol{a}(t) - \boldsymbol{x}(t)))\mu(dt)$.

To demonstrate this, let us first calculate the marginal contribution $v_U(T) - v_U(T \backslash dt)$ of dt to society. Since $v_U(T)$ is achieved at x, trader dt will, after joining the rest of society, consume $\boldsymbol{x}(t)\mu(dt)$, whereas he has contributed $\boldsymbol{a}(t)\mu(dt)$. Therefore his net contribution of commodities to society is $(\boldsymbol{a}(t) - \boldsymbol{x}(t))\mu(dt)$. This is distributed among all other traders, each of whom will receive only an infinitesimal portion of this (in itself infinitesimal) bundle. Therefore the bundle *density* of each trader will change by only an infinitesimal amount, and so by Step 2, the utility of each trader will change by p times the increment in his bundle. Hence the total increment in utility to all members of T due to the distribution of $(\boldsymbol{a}(t) - \boldsymbol{x}(t))\mu(dt)$ among them is $p \cdot (\boldsymbol{a}(t) - \boldsymbol{x}(t))\mu(dt)$. But over and above his contribution of goods to society, dt himself will get utility out of his consumption of $\boldsymbol{x}(t)\mu(dt)$, and this additional utility, which is $u_t(\boldsymbol{x}(t))\mu(dt)$, also must be taken into account when calculating the marginal contribution of dt. We conclude that

$$v_U(T) - v_U(T \backslash dt) = (u_t(\boldsymbol{x}(t)) + p \cdot (\boldsymbol{a}(t) - \boldsymbol{x}(t)))\mu(dt). \qquad (8.2)$$

Now let S be the set of traders up to dt in a random ordering of all the traders. Because of the many traders, the randomly chosen coalition S will almost surely be an almost perfect sample of the population of all traders (insofar as the distribution of utilities and initial bundles is concerned). Hence the contribution $v_U(S)-v_U(S\backslash dt)$ of dt to S is almost surely the same as his marginal contribution $v_U(T)-v_U(T\backslash dt)$ to all of society. Hence the expectation of his contribution to S — which is precisely $(\varphi v_U)(dt)$ — is also the same as his marginal contribution. Step 5 thus follows from (8.2).

Step 6: *Every value allocation is competitive.*

A value allocation $\boldsymbol{x}$ necessarily satisfies (8.1), since by (5.3) and the efficiency of the value (3.1) we have $\int_T u_t(\boldsymbol{x}(t))\mu(dt) = (\varphi v_U)(T) = v_U(T)$. From (5.4) and Step 5 we then obtain $p\cdot(\boldsymbol{a}(t) - \boldsymbol{x}(t)) = 0$ for all t. Thus $p \cdot \boldsymbol{a}(t) = p \cdot \boldsymbol{x}(t)$, and so from Step 4 it follows that $(p, \boldsymbol{x})$ is a competitive equilibrium.

Step 6 is identical to Proposition 5.1, and thus the demonstration of the latter is complete.

Step 7: *If the preferences have concave utilities, then every competitive allocation is a value allocation.*

Let $\{w_t)$ be a family of concave utilities, and let $(p, \boldsymbol{x})$ be a competitive equilibrium. From the concavity it follows that for all t and x, $w_t(\boldsymbol{x}(t)) - w_t(x) \geqq w_t'(\boldsymbol{x}(t)) \cdot (\boldsymbol{x}(t) - x)$, and hence, as in Step 4 above, that the hyperplane through $\boldsymbol{x}(t)$ orthogonal to $w_t'(\boldsymbol{x}(t))$ supports the indifference surface of u_t through $\boldsymbol{x}(t)$. Hence $w_t'(\boldsymbol{x}(t))$ must be proportional to p. So if we define a family $U = \{u_t\}$ of different utilities for the same preferences by multiplying each w_t by an appropriate constant (which may depend on t), then we obtain $p = u_t'(\boldsymbol{x}(t))$ for all t. Hence $u_t(\boldsymbol{x}(t)) - u_t(x) \geqq p \cdot (\boldsymbol{x}(t) - x)$ for all t. From this it follows that if $\boldsymbol{y}$ is another allocation, then $\int_T u_t(\boldsymbol{x}(t))\mu(dt) - \int_T u_t(\boldsymbol{y}(t))\mu(dt) \geqq p \cdot \int(\boldsymbol{x} - \boldsymbol{y}) = p \cdot \int(\boldsymbol{a} - \boldsymbol{a}) = 0$, so that $v_U(T)$ is achieved at $\boldsymbol{x}$, i.e., (8.1) holds. Hence, by Step 5, $(\varphi v_U)(dt) = (u_t(\boldsymbol{x}(t)) + p\cdot(\boldsymbol{a}(t) - \boldsymbol{x}(t)))\mu(dt)$. Since $(p, \boldsymbol{x})$ is a competitive equilibrium, $p \cdot \boldsymbol{a}(t) = p \cdot \boldsymbol{x}(t)$, and hence $(\varphi v_U)(dt) = (u_t \boldsymbol{x}(t))\mu(dt)$. But this is precisely the definition (5.2) of value allocation, and so the proof of Step 7 is complete.

Step 8: *Uniformly smooth preferences can be represented by concave utilities.*

For a time it was thought likely that all preferences with quasi-concave utilities could be represented also by concave utilities; that this is not the case was shown by examples of de Finetti [**7**] and Fenchel [**12**], which will be considered in more detail in Section 5.9 below. At this point we cannot "demonstrate" this step, but perhaps this matter will be clearer to the reader after he has read the discussion in Section 5.9.

We add that in independent work, Mas-Collel [**17**] and Kannai [**16**] have shown that all preferences having quasi-concave utilities can be approximated by preferences having concave utilities, so that Step 8 is "approximately" true even without the smoothness condition.

Proposition 5.2 is, of course, an immediate consequence of Steps 7 and 8. Since Proposition 5.1 has already been established (see Step 6), the demonstration of the Value Equivalence Theorem is complete. The most crucial steps in this demonstration are probably 5 and 8.

5.9. Discussion of Smoothness and Some Counter-Examples

This section is devoted to the differentiability and smoothness conditions for individuals; the uniformity conditions will be discussed in the next section.

If the utilities u_t are not differentiable, the TU game v_U will in general not have an asymptotic value,[15] so one cannot even begin to speak of a value allocation. Moreover, though φv_U may exist under certain circumstances even when the utilities are not differentiable,[16] value theory for the game v_U is in this case quite undeveloped, and

[15] See [**5**, Example 33.9], especially the end of the discussion of that example.
[16] See Footnote 15.

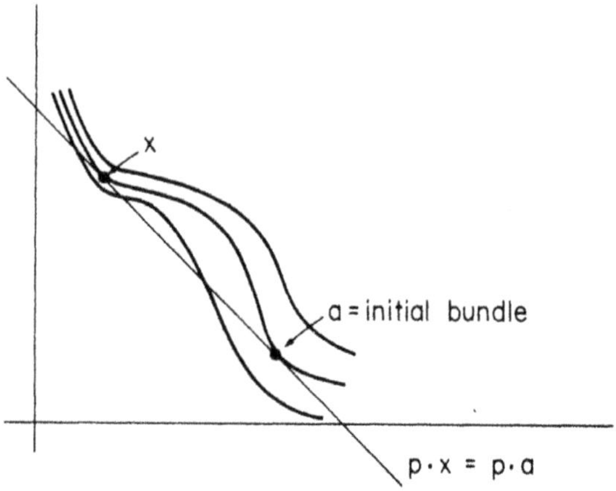

Figure 5.6

we would not be able to apply the basic tools from [**5**] that we need for the proof of our theorem.[17]

So much for the first-order differentiability. Next, we discuss smoothness, which is needed to ensure that the preferences have concave utilities;[18] the importance of this will soon become apparent.

Smoothness has three basic ingredients: quasi-concavity, positive Gaussian curvature, and positive gradients. We discuss the three ingredients separately.

Without quasi-concavity — e.g., for a preference order like that pictured in Figure 5.6 — we will obviously not be able to find concave utilities (the preferred sets $\{x\colon u(x) \geqq y\}$ are convex if u is concave). Moreover, it is then easy to find an example with a competitive allocation that is not a value allocation. Let all traders have the same preferences, and the same initial bundle $\boldsymbol{a}(t) = a$, as pictured in Figure 5.6. Then there is a unique competitive allocation, namely $\boldsymbol{a}$; the prices will of course be the p indicated in Figure 5.6. But

[17]Specifically, Lemma 13.2. On the basis of the available evidence, it seems reasonable to conjecture that without differentiability, something like Proposition 5.1 will still go through, but Proposition 5.2 will not.

[18]More precisely, utilities that are concave over the relevant part of Ω.

now we will argue that $\boldsymbol{a}$ cannot be a value allocation. Indeed, if it were, then there would, in particular, have to exist a bounded differentiable family $U = \{u_t\}$ such that the maximum in the definition of $v_U(T)$ is achieved when $\boldsymbol{x} = \boldsymbol{a}$. By the argument of Step 1 of Section 5.7 (which can easily be made precise), it would follow that $(a, u_t(a))$ would have to be on the boundary of the convex hull of the graph of u_t. If that were the case, there would exist a supporting hyperplane to the graph through $(a, u_t(a))$; and it would necessarily pass through $(x, u_t(x))$. Since the hyperplane supports the graph and passes through both these points on the graph, it follows that $u_t'(a) = u_t'(x)$. But we have constructed the preferences so that though $u_t'(x)$ has the same direction as $u_t'(a)$, it must have a larger magnitude (the indifference curves are closer). Hence $\boldsymbol{a}$ cannot be a value allocation. Indeed, since every value allocation is competitive, and $\boldsymbol{a}$ is the only competitive allocation, there will be no value allocation at all.

The requirement of positive Gaussian curvature says that the indifference surfaces should never be too "flat" — at any point in any direction. Specifically, for x in Ω, consider the hyperplane H_x tangent at x to the indifference surface I_x through x. Let G_x be the line orthogonal to H_x at x (i.e., the line spanned by the gradient at x of a utility function), and let L be any line in H_x through x. We may take G_x and L as coordinate axes in the plane $P_{x,L}$ spanned by them. The situation is pictured in Figure 5.7. Let f be the function (of distance along L) whose graph is the intersection of the indifference surface I_x with the plane $P_{x,L}$. Then the requirement of positive Gaussian curvature says[19] that the second derivative f'' be positive at the origin of $P_{x,L}$; i.e., that the indifference surface not be "too flat" at x in the direction L, and that this be true for all x and all L through x in H_x.

In economic terms, positive Gaussian curvature is equivalent[20] to finite elasticity of substitution.[21] Another familiar characterization is

[19]Given quasi-concavity.
[20]See Footnote 19.
[21]I am indebted for this observation to C. C. von Weiszäcker.

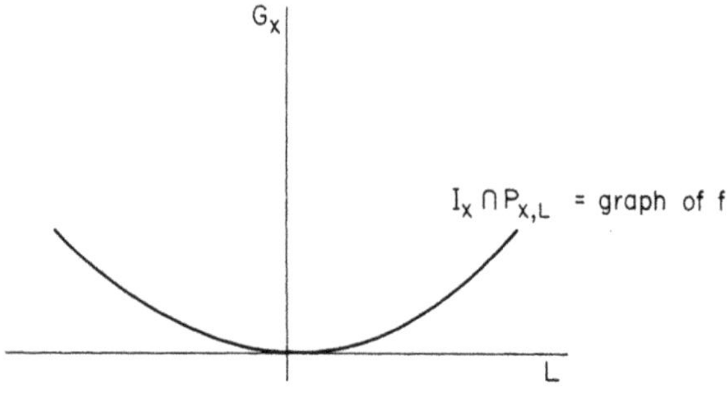

Figure 5.7

that the quadratic form $xu''x^*$ be positive definite when restricted to the plane $u' \cdot x = 0$.

The following example of an otherwise well-behaved preference relation with vanishing Gaussian curvature that has no concave utility is due to W. Fenchel [**12**]. Consider the relation pictured in Figure 5.8, with straight indifference curves that *are not parallel.*[22] If u is any C^1 utility for those preferences, the gradient of u at x has the same direction as the gradient at a; but it has a larger magnitude, since the indifference curves are closer at x. These facts enable us to apply an argument exactly as in our discussion of quasi-concavity, and to deduce that u cannot be concave; and moreover, that in the market M in which all the traders have the preference relations of Figure 5.8 and initial bundles $\boldsymbol{a}(t) = a$, the competitive allocation $\boldsymbol{a}$ is not a value allocation and in fact there is no value allocation.[23]

There is another argument for these facts that is of some interest. Let us think of Figure 5.8 as a contour map, the "height" being the differentiable utility u. The contour lines are closer at x than at y,

---

[22]In Figure 5.8 the indifference curves, when extended, all pass through a single point outside (and to the left of) Ω. But this feature is not crucial; what is needed is only that the indifference curves be straight and closer at x than at y.

[23]In this case there is more than one competitive allocation $\boldsymbol{x}$, since $\boldsymbol{x}(t)$ may be essentially anywhere on the indifference curve of a as long as $\int \boldsymbol{x} = \int \boldsymbol{a}$. However, the argument that *none* of the competitive allocations can be value allocations — and hence that there is no value allocation — is not essentially different from the one we had before.

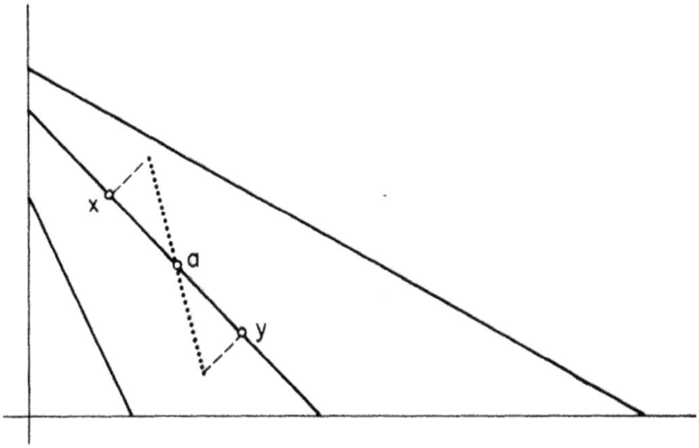

Figure 5.8

which means the ground is steeper there. In particular, the ground rises faster along the dashed line perpendicular to the indifference curve at x than it falls along the parallel dashed line starting at y. So if one were to string a telegraph line along the dotted line and pull it taut, it would pass *over* the indifference curve containing x and y. This means that u cannot be concave. Moreover, in the market M defined in the previous paragraph, our telegraph line and an argument similar to that of Step 1 of Section 5.7 show that $\boldsymbol{a}$ cannot be a value allocation and in fact that there is no value allocation. Moreover, it is fairly clear that this argument goes through even under strict quasi-concavity,[24] as long as the indifference curves are still "very flat" at a, provided, of course, that the tangents to the indifference curves near a are "markedly non-parallel", e.g., that they all pass through the same point (outside of Ω).

We come now to the requirement that the gradients u' be positive. This prevents the "de Finetti phenomenon" [**7**]: the presence of two points x and a such that

$$u(x) = u(a), \quad u'(a) = 0, \quad \text{and} \quad u'(x) > 0 \qquad (9.1)$$

[24]That is, $u(\alpha x + (1-\alpha)y) > \min(u(x), u(y))$ for $\alpha \in (0,1)$.

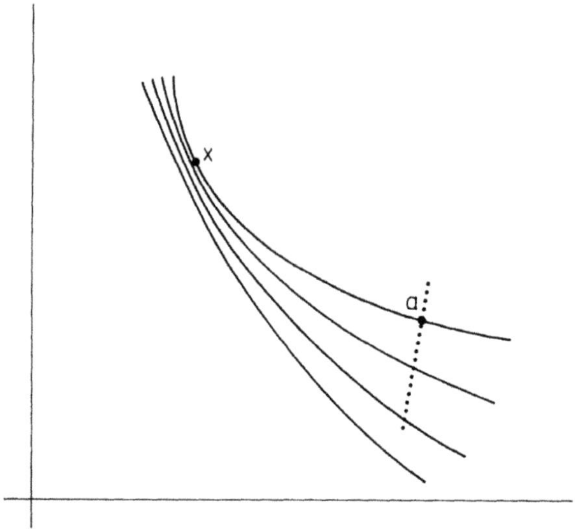

Figure 5.9

(see Figure 5.9). If (9.1) holds for some C^1 utility u, it holds for all such utilities; what it means is that the indifference curves are infinitely more bunched up near x than near a. Under (9.1) u cannot be concave, because that, together with $u'(a) = 0$, would imply that the maximum of u along the line perpendicular to the indifference surface (dotted in the figure) is reached at a; this is impossible if the preferences are monotone.

Consider finally a market, all of whose traders have the same utility obeying (9.1), and the same initial bundle $\boldsymbol{a}(t) = a$. If the utilities are strictly quasi-concave, then the only competitive allocation is $\boldsymbol{x} = \boldsymbol{a}$. If $\boldsymbol{a}$ were a value allocation, there would exist a family $U = \{u_t\}$ of utilities such that $v_U(T)$ is achieved at $\boldsymbol{a}$. This would imply[25] that $(a, u_t(a))$ is on the boundary of the convex hull of the graph of u_t for each t. But since $u_t'(a) = 0$, this supporting hyperplane would be flat, and this would contradict the monotonicity

[25]Step 1 of Section 5.8.

of u_t. Thus $\boldsymbol{a}$ is not a value allocation, and indeed, there is no value allocation.

5.10. Discussion of the Uniformity Conditions

An individual preference relation is *smooth* [**10**] if it has a quasi-concave C^2 utility function with an everywhere strictly positive gradient and indifference curves that have everywhere positive Gaussian curvature. This implies that in each compact subset of Ω, the gradient u' is bounded and bounded away from 0, the Gaussian curvature is bounded away from 0, and the second derivative u'' is bounded. *Uniform* smoothness of a family $\{u_t\}$ requires that the bounds involved in these three statements be independent of t (though they may, of course, depend on the compact subset of Ω). It also requires[26] that the utilities u_t be bounded over all of Ω (i.e., not only compact subsets), uniformly in t.

Of these four requirements, one (about Gaussian curvatures) is stated directly in terms of preferences; but the others are stated in utility terms, and it would be desirable to see what they say about the preferences. The last one — about a uniform bound for the u_t — places no restriction on the preferences; if u_t are utilities satisfying the other restrictions but possibly not uniformly bounded, then $(u_t - u_t(0))/(1 + u_t - u_t(0))$ will satisfy *all* the requirements. The requirement relating to upper and lower positive bounds for the gradient u'_t has two aspects: that relating to the direction and that relating to the magnitude. That relating to the direction has an obvious interpretation in terms of the indifference surfaces (in the two-dimensional case, it simply says that the indifference curves have uniformly bounded slopes). To understand the requirement relating to the magnitudes of the gradients, note that whereas these magnitudes of course depend on the choice of a utility, the *ratio* between the magnitudes at two points x and y on the same

[26]This requirement could be replaced by the far weaker requirement that $u_t(x) = o(||x||)$, integrably in t (see [**4**] or [**5**, Section 31, especially (31.2)]).

indifference surface depends only on the preferences. This ratio is required to be bounded and bounded away from zero.

As for the uniform boundedness of the second derivatives, this is intuitively less clear, but it too can presumably be interpreted in terms of the preferences.

It is perhaps worthwhile to state that a sufficient condition for the family $\{u_t\}$ to be uniformly smooth is that (i) each of the u_t be smooth, (ii) the u_t be bounded over Ω, uniformly in t, and (iii) $\{u_t\}$ be contained in a compact subset of the space $\mathscr{U}$ of C^2 utility functions on Ω, endowed with the topology of C^2 uniform convergence on compact sets.[27]

Assumption 5.2, that $\boldsymbol{a}$ is uniformly bounded, is merely a normalization: if it fails, it is always possible to make it hold simply by replacing the underlying measure μ by the measure v defined by $v(S) = \int_S \Sigma_{i=1}^n \boldsymbol{a}_i(t)\mu(dt)$. In the new representation, all initial bundle densities would be $\leqq (1, \ldots, 1)$. But such a transformation would change also the utilities, and this would lead to complications in the statement of the uniformity conditions. Thus if we wanted to dispense with Assumption 5.2, we would have to replace the fixed compact sets in Conditions 4.2 and 4.3 by sets varying with t in such a way that the maximal distance from the origin is a bounded multiple of $\Sigma_{i=1}^n \boldsymbol{a}^i(t)$. In other words, the "size" of a trader is essentially determined by the size of his initial bundle (not by his measure), and when speaking about bounds of various kinds we must really scale them to the size of the trader. It is in order to avoid these complications that we made the normalization in Assumption 5.2.

5.11. The Theory in a Cardinal Context

The definition of the market model in this paper is an *ordinal* one; that is, it involves only the preferences (and initial bundles) of the

[27] A sub-basis for this topology consists of sets of the form $W(w, \varepsilon, C)$, defined for each w in $\mathscr{U}$, $\varepsilon > 0$, and compact subset C of Ω to be the set of all u in $\mathscr{U}$ such that for all x in C, $|u(x) - w(x)| < \varepsilon$, $\|u'(x) - w'(x)\| < \varepsilon$, and $\|u''(x) - w''(x)\| < \varepsilon$. It is weaker than the Whitney topology discussed in [**27**], so that if a set is compact in the Whitney topology, it is *a fortiori* compact in this topology.

traders, and not any specific choice of utility functions u_t. In the definition of "value allocation", the u_t are not fixed, but are allowed to range over *all* appropriate utility functions that represent the given preferences. Hence the concept of value allocation is also an ordinal one; i.e., whether or not a given $\boldsymbol{x}$ is a value allocation depends only on the preferences (and initial bundles) of the traders.

A purely ordinal model of this kind is, however, not the traditional setting for a value theory. In this section we will consider the Shapley value in a model in which it should feel more at home — one that specifies the *cardinal* utilities of the traders, rather than just their ordinal preferences. By this, we mean a model in which the traders have an intrinsic numerical scale on which the intensity of their preferences can be measured. Thus, it becomes meaningful to say not only that x is preferred to y, and y to z, as in an ordinal context; one can go further and say, for example, that y is relatively close to x on the preference scale (e.g., $\frac{9}{10}$ of the way between z and x). One can lend operational significance to this kind of statement by the method of von Neumann and Morgenstern [**28**], i.e., by referring to lotteries with prizes x, y, and z. But this is by no means the only possible interpretation. For example, cardinal utilities can be defined by the method of Debreu [**9**] in the case of independent commodities, or when the set of commodities can be partitioned into independent "factors" (i.e., sets of commodities); or one can define them by the method of "just noticeable differences" (cf. Fishburn [**13**, p. 18ff. and p. 80ff.]). If one wishes, one can also simply accept intensity of preferences as a fact of life that is not to be ignored, and so assume that cardinal utilities are given *a priori*. In any case, in our application it makes no difference how the cardinal utilities are derived, as long as they are in fact given for all traders.

Formally, call two functions u and v on Ω *equivalent* if there are real numbers α and β, with $\alpha > 0$, such that $u = \alpha v + \beta$; and define a *preference scale* on Ω to be an equivalence class of functions on Ω. A member of the equivalence class is called a *cardinal utility* for that preference scale. Preference scales play the same role in cardinal contexts that preference relations play in ordinal contexts. A cardinal utility is to a preference scale what a utility is to a preference relation; i.e., a representation by means of a numerical function.

It should be particularly noted that though specifying preference scales for all traders enables each trader to compare the intensities of his own preferences, it allows no interpersonal comparisons — no comparison of the intensities of preference of different traders. Our definition of preference scale embodies the familiar statement that "cardinal utilities are determined only up to an additive and a positive multiplicative constant". Thus if we are given cardinal utilities for all traders we can multiply them by arbitrary constants that are different for the different traders, without changing the preference scales, so that no consistent comparisons can be based on such utilities.

To stress the ordinal character of the markets and utilities defined in Section 5.5, we will henceforth in this section refer to them as *ordinal markets* and *ordinal utilities.*

Define a *cardinal market* in exactly the same way as an ordinal market, except that the preference relations $\succ_t$ are replaced by preference scales $\not{p}_t$. Adopt Assumptions 5.1 and 5.2; define *non-atomic*, *allocation*, and *competitive* allocation as in an ordinal market. Define a *cardinal value allocation* to be an allocation $\boldsymbol{x}$ such that there is a bounded differentiable family $U = \{u_t\}$ of cardinal utilities for the preference scales $\not{p}_t$ for which (5.3) holds[28] for all S in $\mathscr{C}$, where the game v_U is defined by (5.1). All this is the same as in Section 5.5, except that we have replaced the preference relations by preference scales.

There are of course relationships between the ordinal and the cardinal concepts. A preference scale $\not{p}$ and a preference relation $\succ$ are called *associated* if the cardinal utilities for $\not{p}$ are ordinal utilities for $\succ$. There is exactly one preference relation associated with each preference scale, but generally many scales with each relation. A cardinal and an ordinal market are *associated* if the ordinal market is obtained from the cardinal market by replacing each of the scales $\not{p}_t$ by the associated relation $\succ_t$. It is easy to see that the value allocations in an ordinal market are precisely those

[28]Or equivalently, (5.2) holds for all t in T.

allocations $\boldsymbol{x}$ that are cardinal value allocations in some associated cardinal market.

As an illustration, consider a cardinal market $\mathscr{M}_c$ with two traders, say $T = \{1, 2\}$. Let u_1 and u_2 be cardinal utilities for $\not{p}_1$ and $\not{p}_2$, and let us assume that they are strictly concave. An *outcome* can be viewed as a pair $(u_1(\boldsymbol{y}(1)), u_2(\boldsymbol{y}(2)))$, where $\boldsymbol{y}$ is an allocation; the set X of all outcomes is compact and convex. There is exactly one cardinal value allocation $\boldsymbol{x}$, and it is such that $(u_1(\boldsymbol{x}(1)), u_2(\boldsymbol{x}(2)))$ is the solution[29] to the "Bargaining Problem" of Nash [**18**], with outcome set X and conflict payoff $(u_1(\boldsymbol{a}(1)), u_2(\boldsymbol{a}(2)))$.

Next, consider the ordinal market $\mathscr{M}_0$ associated with $\mathscr{M}_c$. From an argument of Shapley [**23**, Sec. 3], it follows that *every* Pareto optimal allocation $\boldsymbol{x}$ with $\boldsymbol{x}(1) \succ_1 \boldsymbol{a}(1)$ and $\boldsymbol{x}(2) \succ_2 \boldsymbol{a}(2)$ is a value allocation in $\mathscr{M}_0$; in fact, every such $\boldsymbol{x}$ is the *unique* cardinal value allocation in some cardinal market associated with $\mathscr{M}_0$. Thus, in two-trader markets, the cardinal value allocations depend strongly on the preference scales of the traders, not only on their preference relations; which is of course what one would expect. *A priori*, moreover, there is every reason to expect that the situation is similar also when there are more than two traders, and also in the case of continuous markets. But exactly here we are in for a surprise.

Proposition 11.1. *In a non-atomic cardinal market, assume that there is a bounded differentiable family of concave cardinal utilities for the preference scales. Then the set of cardinal value allocations coincides with the set of competitive allocations.*

The surprising aspect of this is that it asserts equality between the ordinal concept of "competitive allocation" and the cardinal concept of "cardinal value allocation." Non-atomicity turns the concept of "cardinal value allocation" into an ordinal concept — it does not depend on the preference scales, but only on the preference relations. It is to be noted that this is so only when we confine

---

[29]That is, the maximum of $(z_1 - u_1(\boldsymbol{a}(1)))(z_2 - u_2(\boldsymbol{a}(2)))$ over $z \in X$ is achieved when $z = (u_1(\boldsymbol{x}(1)), u_2(\boldsymbol{x}(2)))$.

ourselves to preference scales with concave cardinal utilities; even then, though, it is a startling and somewhat mysterious phenomenon.

On the intuitive, unrigorous level, we can demonstrate Proposition 11.1 by the methods of Section 5.8 — in particular, it follows from Step 6 and the demonstration of Step 7. A rigorous proof will be given in Section 5.14.

Like the Value Equivalence Theorem, Proposition 11.1 (which might be called the "Cardinal Value Equivalence Theorem") depends strongly on the non-atomicity: for finite markets, it is false in both directions. Indeed, we can easily find a two-trader market $\mathscr{M}_c$ of the type discussed above which has a unique competitive allocation and a unique cardinal value allocation, which are different from each other.

The ordinal market $\mathscr{M}_0$ associated with such an $\mathscr{M}_c$ also shows that the Value Equivalence Theorem is false in finite markets, since in $\mathscr{M}_0$ there are many value allocations that are not competitive. It is also false in the converse direction: there are finite markets with competitive allocations that are not value allocations. For an example, take $T = \{1, 2, 3\}$, $n = 2$, $u_t(x) = \sqrt{x^1 + 1} + \sqrt{x^2 + 1}$ for all t, $\boldsymbol{a}(1) = (2, 0)$, $\boldsymbol{a}(2) = (1, 1)$, $\boldsymbol{a}(3) = (0, 2)$. There is a unique competitive allocation $\boldsymbol{x}$, given by $\boldsymbol{x}(1) = \boldsymbol{x}(2) = \boldsymbol{x}(3) = (1, 1)$. Note that $\boldsymbol{x}(2) = \boldsymbol{a}(2)$, so that at $\boldsymbol{x}$, trader 2 gets only his "personal minimum" (what he can guarantee to himself without trading). But any value allocation must yield trader 2 more than his personal minimum, since for any choice of utility functions, he contributes at least his personal minimum in all orderings of the players, and contributes more than his personal minimum in the orderings $(1, 2, 3)$ and $(3, 2, 1)$. Therefore $\boldsymbol{x}$ is not a value allocation.

In game theoretic terms, a cardinal market is an example of a *non-transferable utility (NTU) game*,[30] or what is more commonly called a "game without side payments" (see [**2** or **3**]). Shapley [**23**]

[30]NTU games are generally defined in terms of a specific choice of cardinal utilities, not in terms of preference scales. However, since all concepts associated with such games are usually required to be "invariant" under monotonic linear transformations of the utilities, we are essentially dealing with games defined in terms of preference scales.

has adapted the value notion to NTU games; our definition of "cardinal value allocation" is essentially a specialization to markets of Shapley's NTU value. There is absolutely no reason to expect the NTU value to be invariant under monotonic concave (but non-linear) transformations of the utilities; but Proposition 11.1 asserts that this is precisely what happens in a large class of NTU games, namely the non-atomic cardinal markets. It appears that the game theoretic condition underlying this phenomenon is the "homogeneity" of non-atomic markets: the payoffs available to the members of a coalition depend only on its composition, not on its size; and one may vary the size of a coalition at will, almost without affecting its composition.[31]

5.12. Discussion of the Literature

Though the material of this paper has many roots in previous work, we will here only touch on some of the highlights. Perhaps the first general theorem connecting the value and the competitive equilibrium was established by Shapley [**22**], using the so-called "transferable utility (TU) markets"[32] in an asymptotic model with a fixed finite number of types of traders, in which the number of traders of each type is allowed to tend to ∞. In a TU market with concave differentiable utilities there is a unique competitive payoff,[33] and Shapley proved that the values of the TU games corresponding to the finite markets converge to this payoff. See also Shapley and Shubik [**24**] for a discussion of this work.

Recently, Aumann and Shapley [**5**] studied the value problem in the context of a TU non-atomic market. Assuming differentiability but not concavity, they showed that a TU non-atomic market has a unique competitive payoff, that the corresponding TU game has a

[31]Cf. [**5**, Chapter V]. A more familiar term for homogeneity is "constant returns to scale," as applied to coalitions rather than production processes; increasing or decreasing the scale of a coalition will not affect the possibilities open to its members.

[32]Cf. Footnote 9.

[33]That is, payoff corresponding to a competitive equilibrium.

value, and that the two coincide. No assumption of a finite number of types was used.

In a parallel development Champsaur [**6**] studied the value problem in an asymptotic finite-type framework, but using general Walrasian markets (the kind studied here) rather than TU markets. He considered a special class of cardinal value allocations, namely those obtained by assigning to traders of the same type not only the same preference scale, but actually the same cardinal utility.[34] In that case, traders of the same type get the same utility payoffs; thus if m is the number of types and k the number of traders of each type, then for each k we get a vector $w_k \in E^m$ representing the utility payoffs of the various types. Champsaur [**6**] showed that as $k \to \infty$, every limit point of $\{w_k\}$ must be a competitive payoff. However, he did not prove the converse, that every competitive payoff is such a limit point. Concavity was assumed, but not differentiability.

Thus the situation can be summed up in Table 5.1.

Finally, we mention the important paper of Negishi [**19**], who was the first to use TU markets to prove a theorem about NTU markets;

Table 5.1

	TU	Walrasian
Asymptotic, finitely many types	Shapley [**22**]	Champsaur [**6**]
Non-atomic continuum	Aumann and Shapley [**5**]	This paper

[34]No such assumption is made in our treatment. Every trader may be of a different type; and even if there are many traders of the same type, they may be assigned different cardinal utilities, and in the ordinal case, even different preference scales. Nevertheless, there is implicit in our model a situation that is in principle similar to that of Champsaur. No matter how one assigns the cardinal utilities, one can, because of the non-atomicity, partition practically all the traders (i.e., all but a set of arbitrarily small measure) into a finite number of sets of equal measure, in each of which all the traders will have *approximately* the same cardinal utilities and initial bundles. In some sense, it is this that is responsible for the value equivalence phenomenon. (Compare the discussion of the core equivalence phenomenon at the top of [**1**, p. 49]; there the Debreu-Scarf asymptotic model plays a role similar to that played by the Champsaur model here.)

many of the underlying ideas of the current paper will be seen to be closely related to those of Negishi.

5.13. Proof of Proposition 5.1

We start out by recalling a number of definitions, results, and notations from Aumann–Shapley [**5**].

Let us be given a non-atomic market as described in Section 5.5, and a family $U = \{u_t\}$ of utilities. Define v_U by (5.1).

Lemma 13.1. *If the family U is bounded differentiable, then the maximum in the definition of $v_U(S)$ is attained for all S, and the asymptotic value φv_U of v_U exists.*

Proof. This is an immediate consequence of [**5**, Proposition 31.7], which asserts these conclusions (among others) under conditions on U that are considerably weaker than bounded differentiability.

If $\boldsymbol{x}$ is an allocation, write $u(\boldsymbol{x})$ for the function on T whose value at t is $u_t(\boldsymbol{x}(t))$.

A *transferable utility competitive equilibrium* (TUCE) for U is a pair $(\boldsymbol{x}, p)$, where $\boldsymbol{x}$ is an allocation and $p \in \Omega$, such that for all $t \in T$, $u_t(x) - p \cdot (x - \boldsymbol{a}(t))$ attains its maximum (over $x \in \Omega$) at $x = \boldsymbol{x}(t)$.

Lemma 13.2. *If $(\boldsymbol{x}, p)$ is any TUCE for U, then for all $S \in \mathscr{C}$,*

$$\int_S (u(\boldsymbol{x}) - p \cdot (\boldsymbol{x} - \boldsymbol{a})) = (\varphi v_U)(S).$$

Proof. This is an immediate consequence of [**5**, Proposition 32.3] (see the remark following the statement of the proposition), which asserts this conclusion (as well as some others) under Assumption 5.1 and conditions on U that are considerably weaker than those that define bounded differentiability.

Write $x \geq y$ if $x \geqq y$ and $x \neq y$. The function u_t is called *increasing* if $x \geq y$ implies $u_t(x) > u_t(y)$.

Lemma 13.3. *Let $(\boldsymbol{x}, p)$ be a TUCE for U, such that for almost all t in T, $p \cdot (\boldsymbol{x}(t) - \boldsymbol{a}(t)) = 0$. Assume that the u_t are increasing. Then $\boldsymbol{x}$ is a competitive allocation.*

Proof. From the fact that the u_t are increasing it follows that $p \neq 0$. Since $\boldsymbol{x}$ is by definition an allocation, we must therefore only show that for almost all t, the function u_t is maximized over the budget set $\{x \in \Omega\colon p \cdot x \leqq p \cdot \boldsymbol{a}(t)\}$ by $x = \boldsymbol{x}(t)$. Suppose, indeed, that it were not. Then for a set of t's of positive measure there is a y in Ω such that $u_t(y) > u_t(\boldsymbol{x}(t))$ and $p \cdot y \leqq p \cdot \boldsymbol{a}(t)$. For almost all of those t's we will have $u_t(y) - p \cdot (y - \boldsymbol{a}(t)) \geqq u_t(y) > u_t(\boldsymbol{x}(t)) = u_t(\boldsymbol{x}(t)) - p \cdot (\boldsymbol{x}(t) - \boldsymbol{a}(t))$, in contradiction to the assumption that $(\boldsymbol{x}, p)$ is a TUCE.

Proof of Proposition 5.1. Let $\boldsymbol{x}$ be a value allocation, and let U be the bounded differentiable family of utilities for the market such that for all $S \in \mathscr{C}$,

$$(\varphi v_U)(S) = \int_S u(\boldsymbol{x}). \tag{13.1}$$

In particular, $v_U(T) = (\varphi v_U)(T) = \int_T u(\boldsymbol{x})$, i.e., the maximum in the definition of $v_U(T)$ is attained at $\boldsymbol{x}$. Therefore by [**5**, Proposition 32.1], there is a p such that $(\boldsymbol{x}, p)$ is a TUCE with respect to U. Hence, by Lemma 13.2, for all $S \in \mathscr{C}$ we have

$$(\varphi v_U)(S) = \int_S (u(\boldsymbol{x}) - p \cdot (\boldsymbol{x} - \boldsymbol{a})). \tag{13.2}$$

Combining (13.1) and (13.2), we see that $\int_S p \cdot (\boldsymbol{x} - \boldsymbol{a}) = 0$ for all $S \in \mathscr{C}$. Hence $p \cdot (\boldsymbol{x}(t) - \boldsymbol{a}(t)) = 0$ almost everywhere. But members of a bounded differentiable family are by definition increasing, since $u_t' > 0$. Hence by Lemma 13.3, $\boldsymbol{x}$ is competitive. Thus the proof of Proposition 8.1 is complete.

5.14. Toward the Proof of Proposition 5.2

In this section we prove Proposition 5.2 under Hypothesis (ii), and also Proposition 11.1.

Let $x \in \Omega$. A continuously differentiable function u on Ω is called *concave over* Ω *at* x (or simply *concave at* x) if for all y in Ω, $u(y) - u(x) \leqq u'(x) \cdot (y - x)$. It should be noted that this concept is not a local one; it depends on the behavior of u on all of Ω, not only near x.

Remark 14.1. *A concave continuously differentiable function u on Ω is concave over Ω at each x in Ω.*

Proof. First let x be the interior of Ω. For $0 \leqq \theta \leqq 1$, define $f(\theta) = u(\theta_y + (1-\theta)x)$. Then f is concave and continuously differentiable on $[0, 1]$ and hence $u(y) - u(x) = f(1) - f(0) \leqq f'(0) = u'(x) \cdot (y - x)$. If x is on the boundary of Ω, one obtains the result by a limiting argument from the interior case.

Let b in Ω be a uniform bound on the initial allocation $\boldsymbol{a}$, i.e., $\boldsymbol{a}(t) < b$ for all t; such a bound exists by Assumption 5.2.

For $x \in E^n$, set $\|x\| = \max_i |x^i|$.

Lemma 14.1. *Let $\boldsymbol{x}$ be a competitive allocation, with prices p. Let $\{w_t\}$ be a bounded differentiable family of utilities such that the w_t are concave at each x with $\|x\| \leqq \max_i b^i / \min_i p^i$. Then there is a measurable function $\boldsymbol{\alpha}$ on T, bounded and bounded away from 0, such that for the family $U = \{u_t\}$ of utilities defined by $u_t = \boldsymbol{\alpha}(t) w_t$, we have$(\varphi v_U)(S) = \int_S u(\boldsymbol{x})$ for all $S \in \mathscr{C}$. In particular, $\boldsymbol{x}$ is a value allocation.*

Proof. We will find an $\boldsymbol{\alpha}$ such that $(\boldsymbol{x}, p)$ is a TUCE for U, i.e., such that for almost all t in T, we have that for all y in Ω,

$$u_t(y) - u_t(\boldsymbol{x}(t)) \leqq p \cdot (y - \boldsymbol{x}(t)). \tag{14.1}$$

Since $\boldsymbol{x}$ is a competitive allocation with prices p, and the u_t are increasing, it follows that $p > 0$, and that

$$p \cdot \boldsymbol{a}(t) = p \cdot \boldsymbol{x}(t) \quad \text{for almost all } t. \tag{14.2}$$

Hence for almost all t we have that for all y in Ω,

$$w_t(y) > w_t(\boldsymbol{x}(t)) \Rightarrow p \cdot y > p \cdot \boldsymbol{x}(t). \tag{14.3}$$

For the time being, consider a fixed t satisfying (14.3) for all y in Ω; let $w = w_t$ and $x = \boldsymbol{x}(t)$. Note that for each $i, p^i x^i \leqq p \cdot x = p \cdot \boldsymbol{a}(t) \leqq p \cdot b \leqq \max_i b^i$. Hence

$$\|x\| \leqq \frac{\max_i b^i}{\min_i p^i}, \tag{14.4}$$

and so w is concave at x. Assume first that $x \neq 0$, let $j = \boldsymbol{j}(t)$ be such that $x^j > 0$. Set

$$\alpha = \boldsymbol{\alpha}(t) = \frac{p^j}{w^j(x)}, \tag{14.5}$$

where w^j is the jth derivative of w. We wish to prove (14.1). Without loss of generality assume $j = 1$.

We first show that for all i,

$$p^i \geqq \alpha w^i(x). \tag{14.6}$$

Indeed, suppose $p^i < \alpha w^i(x)$. By (14.5) and $j = 1$, it follows that $p^i/p^1 < w^i(x)/w^1(x)$. Let θ be such that

$$\frac{p^i}{p^1} < \theta < \frac{w^i(x)}{w^1(x)}, \tag{14.7}$$

and consider a point of the form $x_s = x + (-\theta s, 0, \ldots, 0, s, 0, \ldots, 0)$, where the term s occurs in the ith coordinate; if $s > 0$ is sufficiently small, then $x_s \in \Omega$, since $x^1 > 0$. Letting $s \to 0+$, we find $dw(x_s)/ds = -\theta w^1(x) + w^i(x) > 0$, by (14.7). Hence for $s > 0$ sufficiently small, $w(x_s) > w(x)$. But then from (14.3) it follows that $p \cdot x_t \geqq p \cdot x$, i.e., $-p^1\theta s + p^i s > 0$, contradicting (14.7). This proves (14.6).

Next, let i be such that $x^i > 0$. Then we claim that

$$p_i \leqq \alpha w^i(x). \tag{14.8}$$

Indeed, suppose $p^i > \alpha w^i(x)$. Reasoning as in the proof of (14.6), we find a θ such that $(p^i/p^1) > \theta > (w^i(x)/w^1(x))$. Letting $x_s = x + (\theta s, 0, \ldots, 0, -s, 0, \ldots, 0)$, we note that $x_s \in \Omega$ for small s, since $x^i > 0$. Consideration of $dw(x_s)/ds$ then leads to a contradiction as before, so (14.8) is proved.

Let $L = \{i\colon x^i > 0\}$ and $M = \{i\colon x^i = 0\}$. From (14.6) and (14.8) it follows that $p^i = \alpha w^i(x)$ for $i \in L$ and $p^i \geqq \alpha w^i(x)$ for $i \in M$. If $y \in \Omega$, then for $i \in M$ we have $y^i - x^i = y^i \geqq 0$; hence $\alpha w^i(x)(y^i - x^i) \leqq p^i(y^i - x^i)$, with equality holding for $i \in L$. From the concavity of w at f it then follows that $u_t(y) - u_t(x) = \alpha(w(y) - w(x)) \leqq \alpha w'(x) \cdot (y - x) \leqq p \cdot (y - x)$, which is what (14.1) asserts. Thus $\boldsymbol{\alpha}(t)$ is defined and (14.1) is established when $\boldsymbol{x}(t) \neq 0$.

Suppose next that $x = \boldsymbol{x}(t) = 0$. Set $\alpha = \boldsymbol{\alpha}(t) = \min_i p^i / \max_i w^i(0)$. The concavity of w at 0 then gives $u_t(y) - u_t(0) = \alpha(w(y) - w(0)) \leqq \alpha w'(0) \cdot y \leqq p \cdot y$, which is again what (14.1) asserts.

The measurability of $\boldsymbol{\alpha}$ is easily verified. To show that $\boldsymbol{\alpha}$ is bounded and bounded away from 0, note that p is a constant > 0, and w_t' is by assumption bounded and bounded away from 0 in compact sets; in particular, therefore, it is bounded and bounded away from 0 in the compact set

$$\left\{ y \in \Omega\colon \|y\| \leqq \frac{\max_i b^i}{\min_i p^i} \right\},$$

to which, by (14.4), all the $\boldsymbol{x}(t)$ belong.

Thus we have shown that U is bounded differentiable and $(\boldsymbol{x}, p)$ is a TUCE. From Lemma 13.2 and (14.2), we then deduce that for all $S \in \mathscr{C}$, $(\varphi v_U)(S) = \int_S (u(\boldsymbol{x}) - p \cdot (\boldsymbol{x} - \boldsymbol{a})) = \int_S u(\boldsymbol{x})$. This completes the proof of Lemma 14.1.

From Remark 14.1 and Lemma 14.1 it follows that if there exists a bounded differentiable family of concave utilities for the preferences, then every competitive allocation is a value allocation. This yields Proposition 5.2 under Hypothesis (ii), and together with Proposition 5.1 (proved in the previous section), also Proposition 11.1.

5.15. Proof of the Value Equivalence Theorem

In view of Proposition 5.1 and Lemma 14.1, it remains only to prove that if the preferences are uniformly smooth, then we can find a bounded differentiable family of utilities that is concave over Ω in an arbitrarily large set. Specifically, we will prove

Lemma 15.1. *Let $U = \{u_t\}$ be a uniformly smooth family of utilities for the given preferences. Then for every $\gamma > 0$ there exists a bounded differentiable family $\{w_t\}$ of utilities such that the w_t are concave over Ω at each x with $||x|| \leqq \gamma$.*

Remark. Necessary and sufficient conditions for a preference relation to possess a concave utility have been obtained by Fenchel [**12**]. His conditions imply, as a special case, that under appropriate conditions, smooth preferences have concave utilities. Here, however, we have found it more convenient not to use his result, but to prove what is needed "from scratch." This is done for a number of reasons: First, we are interested only in a comparatively simple sufficient condition; whereas the mere statement of Fenchel's necessary and sufficient condition is complex and lengthy. Second, we need a uniform result for a family of preferences, whereas Fenchel's result applies to a single preference relation; therefore even if we used his result we would have to reprove it in the "uniformized" version. Finally, his conditions do not apply exactly here; for example, he considers an open rather than a closed domain of definition. Nevertheless, we will lean heavily on his methods, and there is no question that the principle that "smooth preferences have concave utilities" is to be attributed to Fenchel.

Proof. Let $u_t^i = \partial u_t / \partial x_i$. Since u_t is bounded differentiable, u_t' is uniformly bounded and uniformly positive in compact sets. Hence there is a $\delta > 0$ such that for all t and all x in Ω with $||x|| \leqq \gamma$, we have

$$\min_i \left(\frac{u_t^i}{\sum_{i=1}^n u_t^i} \right) \geqq \delta. \qquad (15.1)$$

Obviously, $\delta \leqq 1/n$. Let $\Gamma = \{x \in \Omega: x \leqq 2\gamma/\delta\}$, and let Δ be the interior of Γ. Set $h(\theta) = -e^{-\xi\theta}$ where $\xi > 0$ is a parameter that will be fixed later. We first wish to show that when ξ is sufficiently large, $h \circ u_t$ is concave on Δ for all t (i.e., $h \circ u_t|\Delta$ is concave for all t). For simplicity of notation, we will for the time being suppress the subscript t, i.e., we will write $u_t = u$.

It is well known (see, for example, [**11**, p. 51]) that a C^2 function w on a convex open set is concave if and only if the matrix $w'' = w''(x)$ of its second partial derivatives is negative semidefinite for each x, i.e., $yw''y^* \leqq 0$ for all y in E^n, where * denotes "transpose." A straightforward calculation yields $(h \circ u)'' = \xi e^{-\xi u}(u'' - \xi u'^* u')$, where u' is a row vector. Thus we must show that $u'' - \xi u'^* u'$ is negative semidefinite for appropriate ξ. We will in fact show that for ξ sufficiently large,

$$u'' - \xi u'^* u' \text{ is negative definite at each } x \text{ in } \Delta, \tag{15.2}$$

i.e., that $y(u'' - \xi u'^* u')y^* < 0$ for all $y \neq 0$.

Let $H (= H_{x,t})$ be the hyperplane $\{y\colon y \cdot u' = 0\}$. Then we assert

$$y \in H \Rightarrow yu''y^* \leqq 0. \tag{15.3}$$

Indeed, for $y \in H$, consider the expression $u(x + \theta y)$ as a function of the real variable θ; it is defined for θ in some neighborhood of 0, since x is in the interior of Ω. The point $x + \theta y$ is in the hyperplane through x orthogonal to u'; by its quasi-concavity, u achieves its maximum in this hyperplane at x. Hence $u(x + \theta y)$, as a function of θ, achieves its maximum at 0, and so

$$0 \geqq \frac{d^2}{d\theta^2} u(x + \theta y) = yu''y^*;$$

this proves (15.3).

Next we assert

$$y \in H \Rightarrow yu''y^* < 0, \quad \text{or } y = 0; \tag{15.4}$$

in other words, not only is u'' negative semidefinite on H, which is what (15.3) asserts, but it is in fact negative definite on H. Suppose it were not; let $y_0 \in H$ be such that $y_0 \neq 0$ and $y_0 u'' y_0^* = 0$

(the possibility $y_0 u'' y_0^* > 0$ is ruled out by (15.3)). By (15.3), $yu''y^*$ attains its maximum subject to $yu'^* = 0$ at $y = y_0$. Therefore there is a Lagrange multiplier λ such that $2u''y_0^* + \lambda u'^* = 0$ and $u'y_0^* = 0$. Hence the system

$$\begin{pmatrix} u'' & u'* \\ u' & 0 \end{pmatrix} z^* = 0$$

of linear equations in $z = (z^1, \ldots, z^{n+1})$ has the distinct solutions $z = 0$ and $z = (y_0, \lambda/2)$, and so the determinant of the matrix defining the system vanishes. But this violates the positivity of the Gaussian curvature (see Condition 4.2 and Equations (4.1) and (4.2)), so (15.4) is proved.[35]

According to a known theorem,[36] (15.4) implies (15.2) for sufficiently large ξ; but we must still show that the choice of ξ may be made independent of x in Δ, and of t.

Let $Q_\xi = u'' - \xi u'^* u'$, let $S^{n-1} = \{y \in E^n\colon ||y|| = 1\}$, and let $g(\xi)$ be the maximum of $yQ_\xi y^*$ over all y in S^{n-1}. Clearly g is non-increasing. Moreover, since $yQ_\xi y^*$ is continuous in y and ξ, and S^{n-1} is compact, it follows that g is continuous. Hence if ξ_0 is the infimum of all ξ such that $g(\xi) < 0$, then $g(\xi_0) = 0$; in other words, $yQ_\xi y^*$ is negative definite for $\xi > \xi_0$, whereas for $\xi = \xi_0$, it is negative semidefinite but not negative definite. Now it is well known that any quadratic form Q can be diagonalized, i.e., there is a non-singular matrix F such that FQF^* is a diagonal matrix; if Q is negative definite then all the diagonal entries are negative, and it follows that the determinant $|Q|$ does not vanish. If Q is negative semidefinite but not definite, then at least one of the diagonal entries vanishes and hence $|Q| = 0$. Thus $|Q_\xi| \neq 0$ for $\xi > \xi_0$ and $|Q_\xi| = 0$ for $\xi = \xi_0$, i.e.,

$$\xi_0 = \max\{\xi\colon |u'' - \xi u'^* u'| = 0\}. \tag{15.5}$$

[35] I am indebted to Kenneth Arrow for this elegant proof of (15.4), which replaces a more awkward proof, using an explicit diagonalization, that I previously had.
[36] For an elegant proof in four lines, see Debreu [**8**, p. 296, Theorem 3].

Now letting i and j run from 1 to n, we have

$$|Q_\xi| = |u'' - \xi u'^* u'| = |u^{ij} - \xi u^i u^j| = \begin{vmatrix} u^{ij} - \xi u^i u^j & u^i \\ 0 & 1 \end{vmatrix}$$

$$= \begin{vmatrix} u^{ij} & u^i \\ u^j & 1 \end{vmatrix} = |u^{ij}| + \xi \begin{vmatrix} u^{ij} & u^i \\ u^j & 0 \end{vmatrix} = |u''| + \xi u^H.$$

From this and (15.5) it follows that $\xi_0 = -|u''|/u^H$. Now making the dependence of u on t explicit once more, and applying (4.2), we obtain

$$\xi_0 = \frac{|u_t''| \; \|u_t'\|_2^{n+1}}{u_t^c}. \tag{15.6}$$

By Conditions 4.2 and 4.3, and the bounded differentiability of the u_t, there is a uniform bound for the right side of (15.6) over all x in Δ and all t. If $\xi > 0$ is chosen greater than that bound, then $u_t'' - \xi u_t'^* u_t'$ is negative definite for all x in Δ and all t, and hence $h \circ u_t$ is concave on Δ. From a simple continuity argument it then follows that $h \circ u_t$ is concave also on the closure Γ of Δ.

It remains to demonstrate the concavity over Ω, for $\|x\| \leqq \gamma$. Let r be a real function of a real variable θ with the following properties:

Property 15.1. *$r \in C^2$ and for all θ, $r'(\theta) > 0$.*

Property 15.2. *$r(\theta) \leqq \min(\theta, 1)$.*

Property 15.3. *$r(\theta) = \theta$ for $\theta \leqq 0$.*

For each t, set $v_t = h \circ u_t$, and define q_t by

$$q_t(\theta) = (\beta_t - \alpha_t) r((\theta - \alpha_t)/(\beta_t - \alpha_t)) + \alpha_t,$$

where $e = (1, \ldots, 1)$, $\beta_t = v_t(2\gamma e)$, and $\alpha_t = v_t(\gamma e)$. Properties 15.1 through 15.3 then yield:

$$q_t \in C^2 \quad \text{and for all } \theta, \quad q_t'(\theta) > 0. \tag{15.7}$$

$$q_t(\theta) \leqq \min(\theta, v_t(2\gamma e)). \tag{15.8}$$

$$q_t(\theta) = \theta \quad \text{for } \theta \leqq v_t(\gamma e). \tag{15.9}$$

Now define $w_t = q_t \circ v_t$. We wish to demonstrate that w_t is concave on $\{x \in \Omega\colon ||x|| \leqq \gamma\}$ over Ω. This means that for all x with $||x|| \leqq \gamma$, and for all y in Ω,

$$w_t(y) \leqq w_t(x) + w_t'(y) \cdot (y - x). \tag{15.10}$$

To prove (15.10), we again suppress the subscript t, writing w for w_t, v for v_t, and u for u_t. First note that because $||x|| \leqq \gamma$, it follows from (15.9) that $w(x) = v(x)$ and $w'(x) = v'(x)$. *If* $y \in \Gamma$, then from (15.8) and the concavity of v on Γ, we get $w(y) = q(v(y)) \leqq v(y) \leqq v(x) + v'(x) \cdot (y - x) = w(x) + w'(x) \cdot (y - x)$, proving (15.10) in this case. If $y \notin \Gamma$, then for at least one coordinate i, we have $y^i \geqq 2\gamma/\delta$. Hence by (15.1),

$$\frac{u'(x)}{\sum_{i=1}^n u^i(x)} \cdot y \geqq 2\gamma = \frac{u'(x)}{\sum_{i=1}^n u^i(x)} \cdot 2\gamma e.$$

Hence $u'(x) \cdot y \geqq u'(x) \cdot 2\gamma e$, and so $v'(x) \cdot y \geqq v'(x) \cdot 2\gamma e$ (since $v'(x) = h'(u(x))u'(x)$ and $h' > 0$ always). Combining this with (15.9) and the concavity of v on Γ we obtain $w(y) = q(v(y)) \leqq v(2\gamma e) \leqq v(x) + v'(x) \cdot (2\gamma e - x) \leqq v(x) + v'(x) \cdot (y - x)$, as was to be shown. Thus (15.10) is established.

Since $\{w_t\}$ is easily seen to be bounded differentiable, the proof of Lemma 15.1 — and with it of the Value Equivalence Theorem — is complete.

Appendix: Marginal and "True" Contributions

In the introduction we stated that "the marginal contribution of a trader to society is almost surely the same as his 'true' contribution." This assertion can be demonstrated on two levels: intuitively, and in a rigorous fashion. The intuitive demonstration was already given in Step 5 of Section 5.7, where we showed that if the coalition S is randomly chosen, then the contribution $v_U(S) - v_U(S \backslash dt)$ of dt to S is almost surely the same as his marginal contribution $v_U(T) -$

$v_U(T \backslash dt)$. We will now state this assertion precisely, and prove it rigorously.

Let us be given a non-atomic market, and let $U = \{u_t\}$ be a bounded differentiable family of utilities for the preferences. Set $v = v_U$. Let $\mathscr{P} = (\Pi_1, \Pi_2, \ldots)$ be a separating sequence of partitions of T into measureable sets, such that each Π_m refines Π_{m-1}. For each m, let $\underset{\sim}{\mathscr{R}}_m$ be a random order[37] on Π_m. For each B in Π_m, denote by $\underset{\sim}{Q} = \underset{\sim}{Q}_B^m$ the union of the members of Π_m up to and including B in the order $\underset{\sim}{\mathscr{R}}_m$. For each S with $B \subset S$ let $\Delta_B v(S) = v(S) - v(S \backslash B)$; $\Delta_B v(S)$ is the *contribution of B to S.*

Proposition A.1. *For every $\varepsilon > 0$ there is an m_0 such that if $m > m_0$ then for every $B \in \Pi_m$ we have*

$$|\Delta_B v(T) - (\varphi v)(B)| < \varepsilon \mu(B) \tag{A.1}$$

and

$$prob\{|\Delta_B v(\underset{\sim}{Q}) - (\varphi v)(B)| < \varepsilon \mu(B)\} > 1 - \varepsilon. \tag{A.2}$$

This says that in a random order on a sufficiently fine partition Π of the player space, with high probability the incremental worth ("true" contribution) of an element B of Π is $(\varphi v)(B) + o(\mu(B))$, and also that the marginal contribution is $(\varphi v)(B) + o(\mu(B))$. In particular, the "true" and marginal contribution differ by a term that is $o(\mu(B))$. Note that $(\varphi v)(B)$ is in general of the order of magnitude of $\mu(B)$ (see Lemma 13.2).

We will make extensive and free use of the notation, terminology, and results of [**5**]. The reader should note that T in this paper plays the same role as I in [**5**], and that $u_t(x)$ is also denoted $u(x, t)$ in [**5**]; also that it is assumed that $u_t(0) = 0$ for all t, an assumption we can make without loss of generality.

There is one respect in which our notation will differ from that of [**5**]. In [**5**], both random variables and functions of t were denoted

[37] A random variable whose values are orders on Π_m, where each of the $|\Pi_m|!$ possible orders is assigned the same probability.

by bold face letters; no confusion could result, since these two kinds of objects were never considered at the same time. Here confusion might result, and so we shall indicate random variables by a wiggly underline, and functions of t by bold face.

Before proceeding with the proof, we shall introduce some auxiliary definitions that will enable us to avoid unnecessary repetition of tiresome "epsilon-delta" phraseology. Let $F = F_B^m$ be a sentence depending on m and on the choice of an element B of Π_m (an example is Formula (A.2)). We shall say that F *eventually* holds if there is an m_0 such that F_B^m holds for $m > m_0$ and all $B \in \Pi_m$. If $\underset{\sim}{f} = \underset{\sim}{f}{}_B^m$ is a random variable depending on m and the choice of a B in Π_m (for example $v(\underset{\sim}{Q})$), then $\underset{\sim}{f}$ is *eventually probably small* if for each $\varepsilon > 0$, prob $\{|\underset{\sim}{f}| < \varepsilon\} > 1 - \varepsilon$ eventually holds (if $\underset{\sim}{f}$ is multidimensional, replace $|\underset{\sim}{f}|$ by $\|\underset{\sim}{f}\|$). If $\underset{\sim}{f} - \underset{\sim}{g}$ is eventually probably small, we shall say that $\underset{\sim}{f}$ is *eventually probably close* to $\underset{\sim}{g}$. Proposition A.1 may be paraphrased by saying that (A.1) holds and that $(\Delta_B v(\underset{\sim}{Q}) - \Delta_B v(T))/\mu(B)$ is eventually probably small.

A number of lemmas will be required. Let $r = r_m = |\Pi_m|$ and let $\underset{\sim}{h} = \underset{\sim}{h}{}_B^m$ be the number of elements of Π_m included in $\underset{\sim}{Q}$, i.e., the number of elements up to and including B in the order $\underset{\sim}{\mathscr{R}}_m$.

Lemma A.1. *Let $\xi \in NA^+$. Then $\xi(\underset{\sim}{Q})$ is eventually probably close to $(\underset{\sim}{h}/r)\xi(T)$.*

Proof. Set $\delta^m = \max\{\xi(A)\colon A \in \Pi_m\}$. Since $\underset{\sim}{h}$ takes on each integer value between 1 and r with the same probability, we have

$$E\left(\xi(\underset{\sim}{Q}\backslash B) - \frac{\underset{\sim}{h}}{r}\xi(T\backslash B)\right)^2 = \frac{1}{r}\sum_{h=1}^{r} E\left(\left(\xi(\underset{\sim}{Q}\backslash B) - \frac{h}{r}\xi(T\backslash B)\right)^2 \,\Big|\, \underset{\sim}{h} = h\right).$$

Referring to the proof of [**5**, Lemma 18.7], we find that each of the summands in the last sum is $\leqq \delta^m$, and that $\delta^m \to 0$. Hence

$E(\xi(\underset{\sim}{Q}\setminus B)-(h/r)\xi(T\setminus B))^2 \leqq \delta^m \to 0$. Since $\xi(B) \leqq \delta^m \to 0$, the lemma follows.

Lemma A.2. *Let $\xi \in NA^+$, $\xi(T) > 0$. Then for each $\theta > 0$, eventually prob $\{\xi(\underset{\sim}{Q}) > \theta\xi(T)\} > 1-2\theta$.*

Proof. Since $\underset{\sim}{h}$ takes on each integer value between 1 and r with equal probability, it follows that for every $\theta > 0$, prob $\{\underset{\sim}{h}/r > \theta\} \to 1-\theta$ as $m \to \infty$. Lemma A.2 then follows from Lemma A.l.

Lemma A.3. *$u'_{\underset{\sim}{Q}}(\int_{\underset{\sim}{Q}} a)$ is eventually probably close to $u'_T(\int a)$.*

Proof. Let $\varepsilon > 0$ be given. Set $\alpha = \sum \int \boldsymbol{a}$. By [**5**, Proposition 38.14], there is a $\delta > 0$ such that for all S with $\mu(S) \geqq \varepsilon$, all b in $A(\varepsilon, \alpha)$, and all δ-approximations $\hat{u}$ to u, we have

$$||\hat{u}'_s(b) - u'_s(b)|| < \varepsilon. \tag{A.3}$$

Let $\hat{u}$ be a δ-approximation to u that is of finite type [**5**, Proposition 35.6], Define an n-dimensional vector measure ζ by $\zeta(S) = \int_S \boldsymbol{a}$. Then

$$\left\| u'_{\underset{\sim}{Q}}\left(\int_{\underset{\sim}{Q}} \boldsymbol{a}\right) - u'_T\left(\int \boldsymbol{a}\right)\right\| \tag{A.4}$$

$$\leqq \|u'_{\underset{\sim}{Q}}(\zeta(\underset{\sim}{Q})) - \hat{u}'_{\underset{\sim}{Q}}(\zeta(\underset{\sim}{Q}))\| + \|\hat{u}'_{\underset{\sim}{Q}}(\zeta(\underset{\sim}{Q})) - \hat{u}'_T(\zeta(T))\|$$

$$+ \|\hat{u}'_T(\zeta(T)) - \hat{u}'_T(\zeta(T))\| = \underset{\sim}{D}_1 + \underset{\sim}{D}_2 + D_3.$$

Applying Lemma A.2 (with $\xi = \mu$ and $\theta = \varepsilon$) and (A.3), we find that eventually

$$\text{prob}\,\{\underset{\sim}{D}_1 < \varepsilon\} > 1 - 2\varepsilon \tag{A.5}$$

and

$$D_3 < \varepsilon. \tag{A.6}$$

To estimate $\underset{\sim}{D}_2$, let k be the number of different "types" in $\hat{u}$, i.e., the number of different $\hat{u}_t$. Denote the nonnegative orthant of

E^k by Ξ. Then there is a continuous function g on $\Xi \times \Omega$ and a k-dimensional vector η of NA^+ measures on T, such that

$$\hat{u}_S(z) = g(\eta(S), z) \tag{A.7}$$

for all $S \in \mathscr{C}$ and $z \in \Omega$, $\eta(T) > 0$, and g is homogeneous of degree 1 on $\Xi \times \Omega$ and continuously differentiable in the interior of $\Xi \times \Omega$ [**5**, (39.17), Propositions 39.8 and 39.13, and Lemma 39.9]. Set $q = n+k$, $\varepsilon_1 = \varepsilon/3q$, and $v = (\eta, \zeta)$. Since $\varepsilon_1 v(T) > 0$, we may find a compact neighborhood U of the straight line segment connecting $\varepsilon_1 v(T)$ to $v(T)$, such that g' is uniformly continuous in U. Hence there is a δ_1, with $0 < \delta_1 < \varepsilon_1$, such that

$$x, y \in U \quad \text{and} \quad \|x - y\| < \delta_1 \Rightarrow \|g'(x) - g'(y)\| < \varepsilon. \tag{A.8}$$

Now by Lemma A.2, eventually prob$\{v(\underset{\sim}{Q}) > \varepsilon_1 v(T)\} > 1 - 2q\varepsilon_1$, and by Lemma A.1, $v(\underset{\sim}{Q})$ is eventually probably close to $(\underset{\sim}{h}/r)v(T)$. Hence by (A.8), eventually prob$\{\|g'(v(\underset{\sim}{Q})) - g'(\underset{\sim}{h}/rv(T))\| < \varepsilon\} > 1 - 3q\varepsilon_1 = 1 - \varepsilon$. But $g'((\underset{\sim}{h}/r)v(T)) = g'(v(T))$, because g is homogeneous. Hence by (A.7), eventually prob$\{\|\hat{u}'_{\underset{\sim}{Q}}(\zeta(\underset{\sim}{Q})) - \hat{u}'_T(\zeta(T))\| < \varepsilon\} > 1 - \varepsilon$. Combining this with (A.5) and (A.6), we find that eventually prob$\{\underset{\sim}{D}_1 + \underset{\sim}{D}_2 + D_3 < 3\varepsilon\} > 1 - 3\varepsilon$, and hence by (A.4), the proof of Lemma A.3 is complete.

Lemma A.4. *For each $\varepsilon > 0$ there is a bounded function $\boldsymbol{\eta}$ such that if $\mu(S) \geqq \varepsilon$, b in Ω satisfies $\Sigma b \leqq \Sigma \int \boldsymbol{a}$, and $u_S(b)$ is attained at $\boldsymbol{y}$, then $\boldsymbol{y}(s) \leqq \boldsymbol{\eta}(s)e$ for almost all $s \in S$.*

Proof. First note that in [**5**, Lemma 37.1] we may choose $\boldsymbol{\eta}_\delta$ bounded; indeed, if the uniform bound on u is γ, then we may choose $\boldsymbol{\eta}_\delta(s) \equiv \max(\gamma/\delta, 1/\delta^2)$. Lemma A.4 then follows immediately from [**5**, Lemma 37.8], when one refers to the first paragraph of the proof of that lemma.

Lemma A.5. *Let $\xi \in NA^+$. Then max$\{\xi(A)\colon A \in \Pi_m\} \to 0$ as $m \to \infty$.*

Proof. This follows from the fact that $\mathscr{P}$ is separating. For details, refer to the proof of [**5**, Lemma 18.7].

Corollary A.1. *Let $v(\underset{\sim}{Q})$ be achieved at $\underset{\sim}{y}$. Then $\int_B \underset{\sim}{y}$ is eventually probably small.*

Proof. Let $\varepsilon > 0$ be given, and let $\boldsymbol{\eta}$ correspond to ε in accordance with Lemma A.4. Let γ be a common bound for $\boldsymbol{\eta}$ and for $\Sigma \boldsymbol{a}$ (see Assumption 5.2). If $\mu(\underset{\sim}{Q}) \leqq \varepsilon$, then $\int_B \underset{\sim}{y} \leqq \int_{\underset{\sim}{Q}} \underset{\sim}{y} = \int_{\underset{\sim}{Q}} \boldsymbol{a} \leqq \gamma\varepsilon e$. If $\mu(\underset{\sim}{Q}) \geqq \varepsilon$, then by Lemma 1.4, $\int_B \underset{\sim}{y} \leqq \gamma\mu(B)e$. But by Lemma A.5, $\mu(B)$ is eventually smaller than ε. Thus the Corollary is proved.

Proof of Proposition A.1. First we demonstrate that eventually (A.1) holds. Suppose $v(T)$ is attained at $\boldsymbol{x}$. From [**5**, Propositions 32.3 and 38.5] it follows that

$$(\varphi v)(B) = \int_B \left[u(\boldsymbol{x}) + u'_T\left(\int \boldsymbol{a}\right)(\boldsymbol{a} - \boldsymbol{x})\right]. \tag{A.9}$$

From [**5**, Lemma 40.1], it follows that

$$\Delta_B v(T) = v(T) - v(T \backslash B) = \int_B [u(\boldsymbol{x}) + u'_{T\backslash B}(c)(\boldsymbol{a} - \boldsymbol{x})], \tag{A.10}$$

where c is on the straight line segment connecting $\int_{T\backslash B} \boldsymbol{x}$ to $\int_{T\backslash B} \boldsymbol{a}$, i.e., (since $\int \boldsymbol{x} = \int \boldsymbol{a}$) $c = \int \boldsymbol{a} - \int_B(\theta\boldsymbol{x} + (1-\theta)\boldsymbol{a})$, where $0 \leqq \theta \leqq 1$. Set $b = \int \boldsymbol{a} - \int_B \boldsymbol{x}$. Then $b - c = (1-\theta)\int_B(\boldsymbol{a} - \boldsymbol{x})$, and hence from [**5**, Proposition 38.7] and Lemma A.5, it follows that for each $\varepsilon_1 > 0$, eventually

$$||u'_{T\backslash B}(b) - u'_{T\backslash B}(c)|| < \varepsilon_1. \tag{A. 11}$$

Again using [**5**, Lemma 40.1], we find

$$u'_{T\backslash B}(b) = u'_T\left(\int \boldsymbol{a}\right), \tag{A.12}$$

on condition that $\int_{T\backslash B} \boldsymbol{a} > 0$; but this follows from Assumption 5.1 and Lemma A.5 for sufficiently large m, so that (A.12) eventually holds. Formulas (A.9), (A. 10), (A.11), and (A.12) yield eventually $|\Delta_B v(T) - (\varphi v)(B)| \leqq \varepsilon_1 \Sigma \int_B (\boldsymbol{a}+\boldsymbol{x})$; together with Assumption 5.2 and Lemma A.4, this yields that (A.1) holds eventually.

The proof of (A.2) is quite similar. Suppose $v(\underset{\sim}{Q})$ is achieved at $\boldsymbol{y}$; [**5**, Lemma 40.1] yields

$$\Delta_B v(\underset{\sim}{Q}) = v(\underset{\sim}{Q}) - v(\underset{\sim}{Q} \backslash B) = \int_B [u(\underset{\sim}{\boldsymbol{y}}) + u'_{\underset{\sim}{Q}}(\underset{\sim}{c})(\boldsymbol{a} - \underset{\sim}{\boldsymbol{y}})], \quad \text{(A.13)}$$

where $\underset{\sim}{c} = \int_{\underset{\sim}{Q}} \boldsymbol{a} + \underset{\sim}{\theta} \int_B (\boldsymbol{a} - \underset{\sim}{\boldsymbol{y}})$ and $0 \leqq \underset{\sim}{\theta} \leqq 1$. Set $\underset{\sim}{b} = \int_{\underset{\sim}{Q}} \boldsymbol{a} - \int_B \underset{\sim}{\boldsymbol{y}}$. Then $\boldsymbol{b} - \boldsymbol{c} = (1-\theta)\int_B (\boldsymbol{a} - \underset{\sim}{\boldsymbol{y}})$, and hence from [**5**, Proposition 38.7], Corollary A.1, and Lemma A.2 (with $\xi = \mu$ and θ small), it follows that $u'_{\underset{\sim}{Q}|B}(\underset{\sim}{b})$ is eventually probably close to $u'_{\underset{\sim}{Q}|B}(\underset{\sim}{c})$. By [**5**, Lemma 40.1], $u'_{\underset{\sim}{Q}|B}(\underset{\sim}{b}) = u'_{\underset{\sim}{Q}}(\int_{\underset{\sim}{Q}} \boldsymbol{a})$ whenever $\int_{\underset{\sim}{Q}\backslash B} \boldsymbol{a} > 0$. By Lemma A.2, for each $\varepsilon_2 > 0$, eventually prob$\{\int_{\underset{\sim}{Q}\backslash B} \boldsymbol{a} > 0\} > 1 - \varepsilon_2$. Hence $u'_{\underset{\sim}{Q}\backslash B}(b)$ is eventually probably close to $u'_{\underset{\sim}{Q}}(\int_{\underset{\sim}{Q}} \boldsymbol{a})$, and by Lemma A.3, this is eventually close to $u'_T(\int \boldsymbol{a})$. Hence

$$u_{\underset{\sim}{Q}}\left(\int_{\underset{\sim}{Q}} \boldsymbol{a}\right) \textit{ and } u'_{\underset{\sim}{Q}\backslash B}(\underset{\sim}{c}) \textit{ are both eventually probably close to } u'_T\left(\int \boldsymbol{a}\right). \quad \text{(A.14)}$$

Now by [**5**, Proposition 38.5], we have $u_t(\underset{\sim}{\boldsymbol{y}}(t)) - u_t(\boldsymbol{x}(t)) \leqq u'_T(\int \boldsymbol{a})(\underset{\sim}{\boldsymbol{y}}(t) - \boldsymbol{x}(t))$. Hence by (A.9) and (A.13),

$$\Delta_B v(\underset{\sim}{Q}) - (\varphi v)(B) \leqq \left(u'_{\underset{\sim}{Q}}(\underset{\sim}{c}) - u'_T\left(\int \boldsymbol{a}\right)\right)\int_B (\boldsymbol{a} - \underset{\sim}{\boldsymbol{y}}). \quad \text{(A.15)}$$

Again by [**5**, Proposition 38.5], $u_t(\boldsymbol{x}(t)) - u_t(\underset{\sim}{\boldsymbol{y}}(t)) \leqq u'_{\underset{\sim}{Q}}(\int_{\underset{\sim}{Q}} \boldsymbol{a})(\boldsymbol{x}(t) - \underset{\sim}{\boldsymbol{y}}(t))$, and hence by (A.9) and (A.13)

$$(\varphi v)(B) - \Delta_B v(\underset{\sim}{Q}) \leqq \left(u'_{\underset{\sim}{Q}}\left(\int_{\underset{\sim}{Q}} \boldsymbol{a}\right) - u'_T\left(\int \boldsymbol{a}\right)\right)\int_B (\boldsymbol{x} - \underset{\sim}{\boldsymbol{y}}) - \left(u'_T\left(\int \boldsymbol{a}\right) - u_{\underset{\sim}{Q}}(\underset{\sim}{c})\right)\int_B (\underset{\sim}{\boldsymbol{y}} - \boldsymbol{a}). \tag{A.16}$$

Now let $\varepsilon_3 > 0$ be given, and let $\boldsymbol{\eta}$ correspond to ε_3 in accordance with Lemma A.4. Let γ be a common bound for $\boldsymbol{\eta}$ and $\Sigma \boldsymbol{a}$ (see Assumption 5.2). By Lemma A.2, eventually $\mu(\underset{\sim}{Q}) \geqq \varepsilon_3$ with probability $\geqq 1 - 2\varepsilon_3$. Set $\varepsilon_4 = \varepsilon_3/2\gamma m$. Hence by (A.14), (A.15), and (A.16), eventually, with probability $\geqq 1 - 2\varepsilon_3$, $|\Delta_B v(\underset{\sim}{Q}) - (\varphi v)(B)| \leqq 4\varepsilon_4\gamma m\mu(B) = 2\varepsilon_3\mu(B)$. This completes the proof of Proposition A.l.

References

1. Aumann, R.: "Markets with a Continuum of Traders," *Econometrica*, 32 (1964), 39–50. [Chapter 3]
2. ———: "A Survey of Cooperative Games Without Side Payments," in *Essays in Mathematical Economics in Honor of Oskar Morgenstern*, ed. by M. Shubik. Princeton, N.J.: Princeton University Press, 1967, 3–27.
3. Aumann, R., and B. Peleg: "von Neumann-Morgenstern Solutions to Cooperative Games Without Side Payments," *Bulletin of the American Mathematical Society*, 66 (1960), 173–179.
4. Aumann, R., and M. Perles: "A Variational Problem Arising in Economics," *Journal of Mathematical Analysis and Applications*, 11 (1965), 488–503.
5. Aumann, R., and L. Shapley: *Values of Non-Atomic Games.* Princeton, N.J.: Princeton University Press, 1971.
6. Champsaur, P.: "Cooperation vs. Competition," *Journal of Economic Theory*, 11 (1975), 394–417.
7. de Finetti, B.: "Sulle Stratificazioni Convesse," *Annali di Matematica*, 4 (1949), 173–183.
8. Debreu, G.: "Definite and Semi-Definite Quadratic Forms," *Econometrica*, 20 (1952), 295–299.

9. ———: "Topological Methods in Cardinal Utility Theory," in *Mathematical Methods in the Social Sciences*, ed. by K. Arrow, S. Karlin, and P. Suppes. Stanford: Stanford University Press, 1959, 16–26.
10. ———: "Smooth Preferences," *Econometrica*, 40 (1972), 603–616.
11. Eggleston, H. G.: *Convexity*. Cambridge: Cambridge University Press, 1958.
12. Fenchel, W.: "Uber konvexe Funktionen mit vorgeschriebenen Niveaumannigfaltigkeiten," *Mathematische Zeitschrift*, 63 (1956), 496–506.
13. Fishburn, P.: *Utility Theory for Decision Making*, New York: Wiley, 1970.
14. Hildenbrand, W.: *Core and Equilibria of Large Economies*. Princeton, N.J.: Princeton University Press, 1974.
15. Kannai, Y.: "Values of Games with a Continuum of Players," *Israel Journal of Mathematics*, 4 (1966), 54–58.
16. ———: "Approximation of Convex Preferences," *Journal of Mathematical Economics*, 1 (1974), 101–106.
17. Mas-Collel, A.: "Continuous and Smooth Consumers: Approximation Theorems," *Journal of Economic Theory*, 8 (1974), 305–336.
18. Nash, J. F.: "The Bargaining Problem," *Econometrica*, 18 (1950), 152–162.
19. Negishi, T.: "Welfare Economics and Existence of an Equilibrium for a Competitive Economy," *Metroeconomica*, 12 (1960), 92–97.
20. Rawls, J.: *A Theory of Justice*. Cambridge, Mass.: Harvard University Press, 1971.
21. Shapley, L.: "A Value for n-Person Games," in *Contributions to the Theory of Games, Vol. II,* ed. by H. W. Kuhn and A. W. Tucker, Princeton, N.J.: Princeton University Press, 1953, 307–317.
22. ———: "Values of Large Games VII: A General Exchange Economy with Money," Rand RM-4248-PR, 1964.
23. ———: "Utility Comparison and the Theory of Games," in *La Decision*, ed. by G. Th. Guilbaud. Paris: Editions du CNRS, 1969.
24. Shapley, L., and M. Shubik: "Pure Competition, Coalitional Power, and Fair Division," *International Economic Review*, 3 (1969), 337–362.
25. Sheshinski, E.: "An Example of Income Tax Schedules Which are Optimal for the Maximin Criterion," forthcoming in *International Economic Review*.
26. Shitovitz, B.: "Oligopoly in Markets with a Continuum of Traders," *Econometrica*, 41 (1973), 467–501.
27. Smale, S.: "Global Analysis and Economics IIA: Extension of a Theorem of Debreu," *Journal of Mathematical Economics*, 1 (1974), 1–14.
28. von Neumann, J., and O. Morgenstern: *Theory of Games and Economic Behavior*. Princeton, N.J.: Princeton University Press, 1944.
29. Whitney, H.: "Functions Differentiable on Boundaries of Regions," *Annals of Mathematics*, 35 (1934), 482–485.

Chapter 6

Agreeing to Disagree*,†

Two people, 1 and 2, are said to have *common knowledge* of an event E if both know it, 1 knows that 2 knows it, 2 knows that 1 knows it, 1 knows that 2 knows that 1 knows it, and so on.

Theorem. *If two people have the same priors, and their posteriors for an event A are common knowledge, then these posteriors are equal.*

If two people have the same priors, and their posteriors for a given event A are common knowledge, then these posteriors must be equal. This is so even though they may base their posteriors on quite different information. In brief, people with the same priors *cannot agree to disagree.*

We publish this observation with some diffidence, since once one has the appropriate framework, it is mathematically trivial. Intuitively, though, it is not quite obvious; and it is of some interest in areas in which people's beliefs about each other's beliefs are of importance, such as game theory[1] and the economics of information.[2] A "concrete" illustration that may clarify matters (and that may be read at this point) is found at the end of the paper.

*Reprinted from *Annals of Statistics* (1976), 4(6), 1236–1239, with permission from Institute of Mathematical Statistics and Wiley.

†This work was supported by National Science Foundation Grant SOC74-11446 at the Institute for Mathematical Studies in the Social Sciences, Stanford University.

[1]Cf. Harsanyi (1967–1968); also Aumann (1974), especially Section 9j (page 92), in which the question answered here was originally raised.

[2]Cf. e.g., Radner (1968) and (1972); also the review by Grossman and Stiglitz (1976) and the papers quoted there.

The key notion is that of "common knowledge." Call the two people 1 and 2. When we say that an event is "common knowledge," we mean more than just that both 1 and 2 know it; we require also that 1 knows that 2 knows it, 2 knows that 1 knows it, 1 knows that 2 knows that 1 knows it, and so on. For example, if 1 and 2 are both present when the event happens and see each other there, then the event becomes common knowledge. In our case, if 1 and 2 tell each other their posteriors and trust each other, then the posteriors are common knowledge. The result is not true if we merely assume that the persons know each other's posteriors.

Formally, let $(\Omega, \mathscr{B}, p)$ be a probability space, $\mathscr{P}_1$ and $\mathscr{P}_2$ partitions of Ω whose join[3] $\mathscr{P}_1 \vee \mathscr{P}_2$ consists of nonnull events.[4] In the interpretation, $(\Omega, \mathscr{B})$ is the space of states of the world, p the common prior of 1 and 2, and $\mathscr{P}_i$ the information partition of i; that is, if the true state of the world is ω, then i is informed of that element $\mathbf{P}_i(\omega)$ of $\mathscr{P}_i$ that contains ω. Given ω in Ω, an event E is called *common knowledge at* ω if E includes that member of the meet[5] $\mathscr{P}_1 \wedge \mathscr{P}_2$ that contains ω. We will show below that this definition is equivalent to the informal description given above.

Let A be an event, and let $\mathbf{q}_i$ denote the posterior probability $p(A|\mathscr{P}_i)$ of A given i's information; i.e., if $\omega \in \Omega$, then $\mathbf{q}_i(\omega) = p(A \cap \mathbf{P}_i(\omega))/p(\mathbf{P}_i(\omega))$.

Proposition. *Let $\omega \in \Omega$, and let q_1 and q_2 be numbers. If it is common knowledge at ω that $\mathbf{q}_1 = q_1$ and $\mathbf{q}_2 = q_2$, then $q_1 = q_2$.*

Proof. Let P be the member of $\mathscr{P}_1 \wedge \mathscr{P}_2$ that contains ω. Write $P = \bigcup_j P^j$, where the P^j are disjoint members of $\mathscr{P}_1$. Since $\mathbf{q}_1 = q_1$ throughout P, we have $p(A \cap P^j)/p(P^j) = q_1$ for all j; hence $p(A \cap P^j) = q_1 p(P^j)$, and so by summing over j we get $p(A \cap P) = q_1 p(P)$. Similarly $p(A \cap P) = q_2 p(P)$, and so $q_1 = q_2$. This completes the proof.

[3] Coarsest common refinement of $\mathscr{P}_1$ and $\mathscr{P}_2$.

[4] Events whose (prior) probability does not vanish.

[5] Finest common coarsening of $\mathscr{P}_1$ and $\mathscr{P}_2$.

To see that the formal definition of "common knowledge" is equivalent to the informal description, let $\omega \in \Omega$, and call a member ω' of Ω *reachable from* ω if there is a sequence $P^1, P^2, \ldots, P^k$ such that $\omega \in P^1, \omega' \in P^k$, and consecutive P^j intersect and belong alternatively to $\mathscr{P}_1$ and $\mathscr{P}_2$. Suppose now that ω is the true state of the world, $P^1 = \mathbf{P}_1(\omega)$, and E is an event. To say that 1 "knows" E means that E includes P^1. To say that 1 knows that 2 knows E means that E includes all P^2 in $\mathscr{P}_2$ that intersect P^1. To say that 1 knows that 2 knows that 1 knows E means that E includes all P^3 in $\mathscr{P}_1$ that intersect P^2 in $\mathscr{P}_2$ that intersect P^1. And so on. Thus all sentences of the form "i knows that i' knows that i knows $\cdots E$" (where $i' = 3 - i$) are true if and only if E contains all ω' reachable from ω. But the set of all ω' reachable from ω is a member of $\mathscr{P}_1 \wedge \mathscr{P}_2$; so the desired equivalence is established.

The result fails when people merely know each other's posteriors. Suppose Ω has 4 elements $\alpha, \beta, \gamma, \delta$ of equal (prior) probability, $\mathscr{P}_1 = \{\alpha\beta, \gamma\delta\}, \mathscr{P}_2 = \{\alpha\beta\gamma, \delta\}, A = \alpha\delta$, and $\omega = \alpha$. Then 1 knows that $\mathbf{q}_2$ is $\frac{1}{3}$, and 2 knows that $\mathbf{q}_1$ is $\frac{1}{2}$; but 2 thinks that 1 may not know what $\mathbf{q}_2$ is ($\frac{1}{3}$ or 1).

Worthy of note is the implicit assumption that the information partitions $\mathscr{P}_1$ and $\mathscr{P}_2$ are themselves common knowledge. Actually, this constitutes no loss of generality. Included in the full description of a state ω of the world is the manner in which information is imparted to the two persons. This implies that the information sets $\mathbf{P}_1(\omega)$ and $\mathbf{P}_2(\omega)$ are indeed defined unambiguously as functions of ω, and that these functions are known to both players.

Consider next the assumption of equal priors for different people. John Harsanyi (1968) has argued eloquently that differences in subjective probabilities should be traced exclusively to differences in information — that there is no rational basis for people who have always been fed precisely the same information to maintain different subjective probabilities. This, of course, is equivalent to the assumption of equal priors. The result of this paper might be considered evidence against this view, as there are in fact people who respect each other's opinions and nevertheless disagree heartily about subjective probabilities. But this evidence is not conclusive: even people who respect each other's acumen may ascribe to each other errors in calculating posteriors. Of course we do not mean simple arithmetical mistakes, but rather systematic biases such as those discussed by Tversky and Kahneman (1974). In private conversation, Tversky has suggested that people may also be biased because of psychological factors, that may make them disregard information that is unpleasant or does not conform to previously formed notions.

There is a considerable literature about reaching agreement on subjective probabilities; a recent paper is DeGroot (1974), where a bibliography on the subject may be found. A "practical" method is the Delphi technique (see, e.g., Dalkey (1972)). It seems to me that the Harsanyi doctrine is implicit in much of this literature; reconciling subjective probabilities makes sense if it is a question of implicitly exchanging information, but not if we are talking about "innate" differences in priors. The result of this paper might be considered a theoretical foundation for the reconciliation of subjective probabilities.

As an illustration, suppose 1 and 2 have a uniform prior on the parameter of a coin, and let A be the event that the coin will come up H (heads) on the next toss. Suppose that each person is permitted to make one previous toss, and that these tosses come up H and T (tails) respectively. If each one's information consists precisely of the outcome of his toss, then the posteriors for A will be $\frac{2}{3}$ and $\frac{1}{3}$ respectively. If each one then informs the other one of his posterior, then they will both conclude that the previous tosses came up once H and once T, so that both posteriors will be revised to $\frac{1}{2}$.

Suppose now that each person is permitted to make several previous tosses, but that neither one knows how many tosses are allowed the other one. For example, perhaps both make 4 tosses, which come up $HHHT$ for 1, and $HTTT$ for 2. They then inform each other that their posteriors are $\frac{2}{3}$ and $\frac{1}{3}$ respectively. Now these posteriors may result from a single observation, from 4 observations, or from more. Since neither one knows on what observations the other's posterior is based, he may be inclined to give more weight to his own observations. Some revision of posteriors would certainly be called for even in such a case; but it does not seem clear that it would necessarily lead to equal posteriors.

Presumably, such a revision would take into account each person's prior on the number of tosses available to him and to the other person. By assumption these two priors are the same, but each person gets additional private information — namely, the actual number of tosses he is allotted. By use of the prior and the information that the posteriors are, respectively, $\frac{2}{3}$ and $\frac{1}{3}$, new posteriors may be calculated. If the players inform each other of these new posteriors, further revision may be called for. Our result implies that the process of exchanging information on the posteriors for A will continue until these posteriors are equal.

References

Aumann, R. J. (1974). Subjectivity and correlation in randomized strategies. *J. Math. Econom.* **1** 67–96. [Chapter 4]

Dalkey, N. C. (1972). *Studies in the Quality of Life.* Lexington Books, Lexington, Mass.

DeGroot, M. H. (1974). Reaching a consensus. *J. Amer. Statist. Assoc.* **69** 118–121.

Grossman, S. J. and J. E. Stiglitz (1976). Information and competitive price systems. *Amer. Econom. Rev.* **46** 246–253.

Harsanyi, J. (1967–1968). Games of incomplete information played by Bayesian players, Parts I–III. *Management Sci.* **14** 159–182, 320–334, 486–502.

Radner, R. (1968). Competitive equilibrium under uncertainty. *Econometrica* **36** 21–58.

Radner, R. (1972). Existence of equilibrium plans, prices, and price expectations in a sequence of markets. *Econometrica* **40** 289–304.

Tversky, A. and D. Kahneman (1974). Judgment under uncertainty: Heuristics and biases. *Science* **185** 1124–1131.

Chapter 7

Correlated Equilibrium as an Expression of Bayesian Rationality*,[1]

Correlated equilibrium is formulated in a manner that does away with the dichotomy usually perceived between the "Bayesian" and the "game-theoretic" view of the world. From the Bayesian viewpoint, probabilities should be assignable to everything, including the prospect of a player choosing a certain strategy in a certain game. The so-called "game-theoretic" viewpoint holds that probabilities can only be assigned to events not governed by rational decision makers; for the latter, one must substitute an equilibrium (or other game-theoretic) notion. The current formulation synthesizes the two viewpoints: Correlated equilibrium is viewed as the *result* of Bayesian rationality; the equilibrium condition appears as a simple maximization of utility on the part of each player, given his information.

*Reprinted from *Econometrica* (1987), 55(1), 1–18, with permission from Wiley.

[1]Research supported by the Institute for Mathematical Studies in the Social Sciences, Stanford University, under National Science Foundation Grant SES 83-20453; by the Institute for Mathematics and its Applications, University of Minnesota; and by the Mathematical Sciences Research Institute, Berkeley. We are grateful to Lloyd Shapley for an illuminating conversation, a decade back, that clarified for us the equivalence between the two definitions of correlated equilibrium; to Kenneth Arrow, for motivating us to write this paper; to Bob Weber, for the very apt quotation from Burns; and to Dave Kreps and two anonymous referees, for splendid editorial work.

A feature of this approach is that it does not require explicit randomization on the part of the players. Each player always chooses a definite pure strategy, with no attempt to randomize; the probabilistic nature of the strategies reflects the uncertainty of *other* players about his choice. Examples are given.

7.1. Introduction

O wad some pow'r the giftie gie us
To see oursels as ithers see us!
—Burns

The Equilibrium Concept of Nash (1951), together with its refinements,[2] is without doubt the single game-theoretic tool that is most often applied in economics. Yet, though at first the definition seems simple and natural enough, a little reflection leads to some puzzlement as to why and under what conditions the players in an n-person game might be expected to play such an equilibrium. Recall that it is defined as an n-tuple of strategies in which the component strategy of each player maximizes that player's utility given that the other players are playing their components. Now why should any player assume that the other players will play their components of such an n-tuple, and indeed why should they? This is particularly perplexing when, as often happens, there are multiple equilibria; but it has considerable force even when the equilibrium is unique. Indeed, there are games whose Nash equilibria appear quite unattractive even though they are unique (see Harsanyi, 1977, p. 125); and even without such examples, it does not seem clear why the players would play even a unique Nash equilibrium. In a two-person game, for example, Player 1 would play his component only if he believes that 2 will play his; this in turn would be justified only by 2's belief that 1 will indeed play his component; and so on. To many this will sound

[2]E.g., Selten (1975), Myerson (1978), Kreps and Wilson (1982), Kalai and Samet (1984), Kohlberg and Mertens (1986).

like a plain old circular argument: consistent, perhaps, but hardly compelling.

Nash equilibrium does make sense if one starts by assuming that, for some specified reason, each player knows which strategies the other players are using. But this assumption appears rather restrictive.

Another criticism of equilibrium that has been advanced is that it appears inconsistent with the modern subjectivist, Bayesian view of the world. According to the Bayesian view, subjective probabilities should be assignable to every prospect, including that of players choosing certain strategies in certain games. Rather than playing an equilibrium, the players should simply choose strategies that maximize their utilities given their subjective distributions over the other players' strategy choices.

It is the purpose of this paper to provide a simple rationale for equilibrium. Surprisingly, our rationale is precisely in terms of the "criticism" in the previous paragraph. We will show that, far from being inconsistent with the Bayesian view of the world, the notion of equilibrium is an unavoidable consequence of that view. It turns out, though, that the appropriate equilibrium notion is not the ordinary mixed strategy equilibrium of Nash (1951), but the more general notion of correlated equilibrium.

Specifically, we will show the following. Let us be given an n-person game G in strategic (i.e. normal) form. Assume that (i) as in Savage (1954), each player has a subjective probability distribution over the set of all states of the world; and (ii) it is common knowledge[3] that each player chooses a strategy that maximizes his expected utility given his information. Then the strategies chosen by the players constitute a correlated equilibrium of G. Conversely, each correlated equilibrium of G can be obtained in this way for an

[3]An event is *common knowledge* among a set of agents, if it is known to all, it is known to all that it is known to all, and so on ad infinitum. See Lewis (1969) and, for a precise mathematical formulation, Aumann (1976).

appropriate choice of the parameters. A precise formulation and proof will be found in Section 7.3.

To some readers, this may seem obvious. Once one has the formulation, it is indeed embarrassingly easy to prove, as we shall see. But it should not be confused with certain characterizations of Nash equilibrium, which look similar but are actually much more transparent. There one assumes not only that each player is a utility maximizer, but that he knows the strategies actually being used by all other players. We make no such assumption; indeed, in our treatment, the players do *not* in general know how others are playing. We assume only that it is common knowledge that all the players are Bayesian utility maximizers, that they are rational in the sense that each one conforms to the Savage theory. Such an assumption underlies most of game theory, and of economic theory as well; we show that it, by itself, is enough to assure that the outcome is a correlated equilibrium.

An important feature of our approach is that it does not require explicit randomization on the part of the players. Each player always chooses a definite pure strategy, with no attempt to randomize; the probabilistic nature of the strategies reflects the uncertainty of *other* players about his choice. This will be discussed in detail in Section 7.6.

The basic idea of this paper is extraordinarily simple; the main theorem is proved in less than a dozen lines. Some conceptual issues do arise, but still, we cannot grasp how such a simple idea grew into a paper of this length. A reader just interested in the main points, who knows what a correlated equilibrium is, should read Sections 7.3 and 7.6 only. Those not familiar with correlated equilibria should read Section 7.2 first.

7.2. Correlated Equilibrium

Let G be an n-person game in strategic form. For $i = 1, \ldots, n$, denote by S^i the set of pure strategies — henceforth *actions* — of Player i, and by $h^i(s)$ the payoff to i for an n-tuple $s := (s^1, \ldots, s^n)$ of actions.[4] Set

[4]The symbol := means that the expression on the left is being defined.

$S := S^1 \cdots \times S^n$ and $h := (h^1, \ldots, h^n)$. Define a *correlated strategy n-tuple* in G as a function f from a finite[5] probability space Γ into S; that is, as a random variable whose values are n-tuples of actions.

Correlation is a more general form of strategy randomization than mixing. In both cases, the players base their choice of an action on observation of a random event; with mixed strategies, the observations are independent, whereas with correlated strategies, they need not be. To see that mixed strategies may be formally viewed as correlated strategies, take Γ to be the product probability space $\Gamma^1 \times \cdots \times \Gamma^n$, where Γ^i is the probability space by means of which i's mixed strategy is defined — the set of outcomes of his roulette spin. The simplest examples of "true" correlation (nonindependent action choices) occur when all players base their choices on observation of the *same* random variable. For example, in the prisoner's dilemma of Figure 7.1, the symmetric efficient outcome can be achieved if *both* players simultaneously observe a single toss of a fair coin, then play (top, right) or (bottom, left) according to whether heads or tails came up. It cannot, however, be achieved by mixed strategies.

In general, the observations of the players need be neither identical nor independent. Chance chooses an element γ of Γ, then suggests to each player i that he take the action $f^i(\gamma)$ $(:= (f(\gamma)^i))$. If all players follow the suggestions, the correlated strategy n-tuple f results.

The function f may be identified with the n-tuple $(f^1, \ldots, f^n)$. If g^i is a different function from Γ to S^i, then as usual,

$$(f^{-i}, g^i) := (f^1, \ldots, f^{i-1}, g^i, f^{i+1}, \ldots, f^n).$$

2, 2	0, 6
6, 0	1, 1

Figure 7.1

---

[5]I.e., with a finite number of points (and so also of events). The finiteness of Γ plays no role in the sequel; we assume it only to avoid complicating the model with irrelevant generality.

Throughout the sequel, E denotes "expectation." If f is a correlated strategy n-tuple, note that $h^i(f)$ is a real-valued random variable.

Definition 2.1. A *correlated equilibrium* (c.e.) in G is a correlated strategy n-tuple f such that

$$Eh^i(f) \geq Eh^i(f^{-i}, g^i) \tag{2.2}$$

for each player i and each g^i that is a function of f^i.

In form, this definition is very similar to that of Nash equilibrium. Equilibrium is achieved when no player can gain by deviating from the suggestions, given that the others obey them. The deviations g^i are restricted to be functions of f^i (i.e., compositions of some other function with f^i) because i is informed of $f^i(\gamma)$ only, and so can distinguish only between members of Γ that are distinguished by f^i.

It will be useful to define the *distribution* of a correlated strategy n-tuple f as the function that assigns to each n-tuple s of actions the number $\text{Prob}\{f^{-1}(s)\}$. Much like mixed strategies, correlated strategy n-tuples can for most practical purposes be identified with their distributions. Thus the correlated strategy n-tuple discussed above (Figure 7.1) can, using a familiar notation, be written $\frac{1}{2}(\text{top, right}) + \frac{1}{2}(\text{bottom, left})$. Another way to represent the distribution is to insert the appropriate probability in each square of the payoff matrix; for the above example, the distribution is that of Figure 7.2.

This distribution is not a *correlated equilibrium distribution* (c.e.d.), i.e., it does not represent a correlated equilibrium. Indeed, in the prisoner's dilemma, it is *always* worthwhile for Player 1 to play bottom, also if it is suggested that he play top. Figure 7.3 contains an example of a game ("the battle of the sexes") and a correlated

0	$\frac{1}{2}$
$\frac{1}{2}$	0

Figure 7.2

2, 1	0, 0
0, 0	1, 2

$\frac{1}{2}$	0
0	$\frac{1}{2}$

Figure 7.3

6, 6	2, 7
7, 2	0, 0

Figure 7.4

$\frac{1}{3}$	$\frac{1}{3}$
$\frac{1}{3}$	0

Figure 7.5

strategy distribution that *is* in equilibrium. If top is suggested to 1, then he knows that left was suggested to 2, so he would lose by deviating to bottom; similarly if bottom is suggested, and for Player 2.

By similar methods, it may be seen that any convex combination of Nash equilibria is a correlated equilibrium. But there are also other kinds of correlated equilibria. For example, Figure 7.4 illustrates a two-person game ("chicken") with three Nash equilibria, whose payoffs are (2,7), (7,2), and $(4\frac{2}{3}, 4\frac{2}{3})$. The distribution of Figure 7.5 may be seen to be a correlated equilibrium, whose payoff (5, 5) is outside the convex hull of the Nash equilibrium payoffs. It follows that the distribution of Figure 7.5 is itself outside the convex hull of the Nash equilibria. Other examples of correlated equilibria that are outside the convex hull of the Nash equilibria may be found in Aumann (1974); for a particularly beautiful example, see Moulin and Vial (1978).

Computationally, correlated equilibria are simpler objects than Nash equilibria. The correlated equilibrium distributions of a given game G constitute a compact convex polyhedron whose defining linear inequalities can be explicitly written down. Here we treat only the two-person case, which is notationally less cumbersome; the principle, however, is no different in the general case.

Let $n = 2$; set $l := |S^1|$, $m := |S^2|$, $h_{jk} := h(j,k)$ for $j \in S^1$ and $k \in S^2$. Let $\sum_j$ and $\sum_k$ denote summation over all j in S^1 and all k in S^2 respectively. The distribution of a correlated strategy pair is an lm-tuple (p_{jk}), where j and k range over S^1 and S^2 respectively,

$$p_{jk} \geq 0 \quad \text{for all } j \text{ and } k, \quad \text{and} \quad \sum_j \sum_k p_{jk} = 1.$$

We will call such objects simply *distributions* for short.

Proposition 2.3. *A distribution* (p_{jk}) *is a correlated equilibrium distribution if and only if*

$$\sum_k (h^1_{jk} - h^1_{qk}) p_{jk} \geq 0 \text{ for all } j, q \text{ in } S^1, \quad \text{and} \tag{2.4}$$

$$\sum_j (h^2_{jk} - h^2_{jr}) p_{jk} \geq 0 \qquad \text{for all } k, r \text{ in } S^2. \tag{2.5}$$

Proof. For (2.2) it is necessary and sufficient that for each i, the expectation still obeys the inequality when conditioned on each possible value of f^i, i.e., on each possible "suggestion" to Player i (a suggestion is called *possible* if it has positive probability). Suppose j is a possible suggestion to Player 1, i.e., $\sum_k p_{jk} > 0$. Given this suggestion, the conditional probability of 1 that k was suggested to 2 is $p_{jk}/\sum_k p_{jk}$; hence 1's conditional expected payoff is $\sum_k h^1_{jk} p_{jk}/\sum_k p_{jk}$, which we denote $H^1(j|j)$. If 1 were to deviate to the action q, his conditional expected payoff would be $\sum_k h^1_{qk} p_{jk}/\sum_k p_{jk}$, since the payoffs change, but 1's information remains the same; we denote this by $H^1(q|j)$. The equilibrium condition (2.2) for $i = 1$ says that deviation to q is not worthwhile, i.e., that $H^1(j|j) \geq H^1(q|j)$; if we multiply through by $\sum_j p_{jk}$, we obtain (2.4). Similarly, (2.5) is seen to express the equilibrium condition (2.2) for $i = 2$.

Definition 2.1 appears a little different from our 1974 definition of correlated equilibrium; there, the players may get more information about γ than just the suggested action $f^i(\gamma)$. It turns out, though, that practically speaking, the two definitions are equivalent. This will be discussed again below (Section 7.4g).

7.3. Bayesian Rationality in Games

We assume exogenously given an n-person game as in the previous section. Also exogenously given are the following: (i) a finite[6] set Ω, with generic element ω; (ii) for each player i in G, a probability measure p^i on Ω; (iii) for each player i in G, a partition $\mathscr{P}^i$ of Ω. The set Ω represents the set of all possible *states of the world*, and each ω denotes a specific such state. The probability measure p^i is i's *prior* on Ω. The partition $\mathscr{P}^i$ is i's *information partition*; if the true state of the world is $\omega \in P \in \mathscr{P}^i$, then i knows that some element of P is the true state of the world, but he does not know which one it is.

The term "state of the world" implies a definite specification of all parameters that may be the object of uncertainty on the part of any player of G. In particular, each ω includes a specification of which action is chosen by each player of G at that state ω. Conditional on a given ω, everybody knows everything; but in general, nobody knows which is really the true ω. Taking the "atoms" of Ω to represent complete specifications of all possible variables enables us to represent *all* aspects of uncertainty on the part of any player — including uncertainty about the uncertainty of other players — by means of the partitions $\mathscr{P}^i$. This constitutes the standard model for "differential" ("incomplete") information in multi-person contexts; its implications will be discussed in more detail in the next section.

We will use the following assumption:

Common Prior Assumption: All the priors p^i are the same; that is, there is a probability measure p on Ω such that $p^1 = p^2 = \cdots = p^n = p$.

This is sometimes called the *Harsanyi Doctrine.* It does *not* imply that all players have the same subjective probability. The

---

[6]See Section 7.4c.

subjective probability of a player is his posterior given his information; these may well be different. Roughly speaking, the common prior assumption says that differences in probability estimates of distinct individuals should be explained by differences in information and experience. This, too, will be discussed further in the sequel, and we will also indicate how our results must be modified when the common prior assumption is relaxed or abandoned.

Denote by $\mathbf{s}^i(\omega)$ the action chosen by Player i in the game G at the state ω of the world. Formally, each $\mathbf{s}^i$ is simply an exogenously specified function from Ω to S^i; it may be considered item (iv) in the informational model of the world specified at the beginning of this section. We make the modest assumption that each player i knows which action he chooses, i.e., that $\mathbf{s}^i$ is measurable[7] w.r.t. (with respect to) $\mathscr{P}^i$.

Let $\mathbf{s}(\omega) := (\mathbf{s}^i(\omega), \ldots, \mathbf{s}^n(\omega))$ be the n-tuple of actions chosen at state ω. Call a player i *Bayes rational at* ω if his expected payoff given his information, $E(h^i(\mathbf{s})|\mathscr{P}^i)(\omega)$, is at least as great as the[8] amount $E(h^i(\mathbf{s}^{-i}, s^i)|\mathscr{P}^i)(\omega)$ that it would have been had he chosen an action s^i other than the action $\mathbf{s}^i(\omega)$ that he did in fact choose; in brief, if he chooses an action that maximizes his payoff given his information.

Main Theorem: *If each player is Bayes rational at each state of the world, then the distribution of the action n-tuple* $\mathbf{s}$ *is a correlated equilibrium distribution.*

Proof. We must exhibit a probability space Γ and a function $f\colon \Gamma \to S$ that is a correlated equilibrium and that has the same distribution as $\mathbf{s}$. Indeed, if we set $\Gamma := (\Omega, p)$, then $\mathbf{s}$ itself is such a function. To

---

[7]A random variable is *measurable* w.r.t. a partition if it is constant on each element of the partition. Intuitively, this means that the r.v. contains no more information than the partition; a person who knows which event in the partition obtains, also knows the value of the r.v.

[8]Recall that if x is a random variable, then $E(\mathbf{x}|\mathscr{P}^l)$ is defined as the r.v. whose value at a specific ω is $E(\mathbf{x}|P)$, where P is the event in $\mathscr{P}^l$ that contains ω. (P represents i's knowledge, the smallest event that he knows to have occurred.)

see this, let $i \in N$, and let $g^i \colon \Omega \to S^i$ be any function of $\mathbf{s}^i$. Since $\mathbf{s}^i$ is $\mathscr{P}^i$-measurable, g^i also is; that is, it is constant on each P in $\mathscr{P}^i$. The Bayes rationality of i at each state implies that for each such P and each s^i in S^i,

$$E(h^i(\mathbf{s})|P) \geq E(h^i(\mathbf{s}^{-i}, s^i)|P).$$

Taking s^i to be the constant value of g^i throughout P yields

$$E(h^i(\mathbf{s})|P) \geq E(h^i(\mathbf{s}^{-i}, g^i)|P);$$

multiplying both sides by Prob P and summing over all P in $\mathscr{P}^i$ yields

$$E(h^i(\mathbf{s})) \geq Eh^i(\mathbf{s}^{-i}, g^i).$$

For $f = \mathbf{s}$, this is precisely the condition (2.2) that defines correlated equilibrium.

7.4. Discussion

(a) *Personal Choice as a State Variable.* The chief innovation in our model is that it does away with the dichotomy usually perceived between uncertainty about acts of nature and of personal players. Of course, a player always knows which decision he himself takes; here, this information is not treated differently from private information he may have about other aspects of the state of the world.

This point is a little subtle and is worth some discussion. In traditional Bayesian decision theory, each decision maker is permitted to make whatever decision he wishes, after getting whatever information he gets. In our model this appears not to be the case, since the decision taken by each decision maker is part of the description of the state of the world. This sounds like a restriction on the decision maker's freedom of action; at a given state ω, it is as if the model forces him to take the decision dictated by ω.

But closer examination reveals that "freedom of choice" is not an issue. The model describes the point of view of an outside observer. Such an observer has no *a priori* knowledge of what the players will choose; for him, the choices of the players are part of the description of the states of the world. This does not mean that the players cannot choose whatever they want, but only that the observer will not in general know what they want.

This "outside observer" perspective is common to all differential information models in economics (as well as to all extensive games that are not of perfect information). In such models, each player gets some information or "signal"; he hears only the signal sent to him, not that of others. In analyzing this situation, each player i must first look at the whole picture as if he were an outside observer; he cannot ignore the possibility of his having gotten a signal other than the one he actually got, even though he knows that he did not actually get such a signal. This is because the *other* players do not know what signal he got. Player i must take the ignorance of the other players into account when deciding on his own course of action, and he cannot do this if he does not explicitly include in the model signals other than the one he *knows* he got. The "outside observer" referred to above is thus a surrogate for the ignorance of the system as a whole — the lack of common knowledge — of the signals received by each player.

Similarly, each player is of course free to choose whatever action he wishes. That we include his action as part of the description of the state of the world is only a convenient way of expressing the fact that the other players do not know which action he wishes to choose.

The reader may still be puzzled by the fact that in deriving his posterior about other players' choices, each player conditions on his own information, which includes his own choice. How can a player's own choice help him to guess what other players will do?

But again, this paradox is illusory. The player is not really conditioning on his choice, but on the substantive information that leads him to make this choice. This substantive information leads to a posterior on the other players' choices, which in turn leads to an optimal choice, or to a set of such choices. Intuitively, the fact that a player's choice is part of his information is used not so much in deriving his own posterior about what others will do, but rather in estimating the others' posteriors about what he will do. They cannot simply make arbitrary "guesses" about this, but must take into account that he is maximizing given his own information. Since he is reasoning similarly about them, it appears that one might be led to troubling circular reasoning. This precisely is avoided by the model of the previous section.

To sum up, a game in which the players do not know what the other players do should be treated like any other model of differential information, where the possible values of the parameters about which some of the players are ignorant are used as specifications of the state of the world.

(b) *Common Knowledge of Information Partitions and Priors.* A question that often crops up when models of differential information are discussed is whether there can be uncertainty on the part of one player about the information partitions $\mathscr{P}^i$ of other players.

The answer is "no". While Player 1 may well be ignorant of what Player 2 knows — i.e., of the element of $\mathscr{P}^2$ that contains the "true" state ω of the world — 1 cannot be ignorant of the *partition* $\mathscr{P}^2$ itself. Indeed, $\mathscr{P}^2$ is part of the description of the model, and does not enter the description of any particular ω; it therefore cannot be the object of uncertainty, it must be common knowledge.

This is not an assumption, but a "theorem", a tautology; it is implicit in the model itself. Since the specification of each ω includes a complete description of the state of the world, it includes also a list of those other states ω' of the world that are, for Player 2, indistinguishable from ω. If there were uncertainty about this list on the part of Player 1 (or of any other player), then the description of ω would not be complete; one should then split ω into several states, depending on which states are, for 2, indistinguishable from ω. Therefore the very description of the ω's implies the structure of $\mathscr{P}^2$, and similarly of all the $\mathscr{P}^i$. The description of the ω's involves no "real" knowledge; it is only a kind of code book or dictionary. The structure of the $\mathscr{P}^i$ also involves no real knowledge; it simply represents different methods of classification in the dictionary.

The situation with priors is similar. Once one accepts the Bayesian viewpoint, that each player has a prior on Ω, it follows that there cannot be any uncertainty on the part of one player about other players' priors. Each player's prior must be common knowledge among all players. (This in itself does not imply the further assumption, which we have made, that all the priors are the same; this further assumption will be discussed separately below.)

The reasoning leading to the conclusion that the priors are common knowledge is similar to that leading to the conclusion that the $\mathscr{P}^i$ are common knowledge. It is enough to convince ourselves that Player 1's prior $p^1(\omega)$ for each given state ω of the world is common knowledge. If it were not, then the description of ω would be incomplete; we would be able to split ω into several states, depending on the various possibilities for $p^1(\omega)$.

The reader may ask why we adduce verbal, informal arguments for these assertions, rather than proving them as formal theorems. The answer is that the assertions have no formal content. Within the model, knowledge refers to events, i.e., sets of states. One can ask whether at a given state ω, a player i knows an event A; this is the case if and only if A includes that element of $\mathscr{P}^i$ that contains ω. But neither a partition nor a prior is an event; formally speaking, the concepts of knowledge and common knowledge do not apply to them. In the above informal discussion, when we refer to the partitions, say, as "common knowledge", we mean that all players know them, all know that all know them, and so on, where "know" has its informal, everyday meaning.

It is important to realize that while the above assertions (that the partitions and priors are common knowledge) aid us in understanding the model, they do not affect the conclusions. Each player uses only his own prior and his own partition in reaching a decision; within the model, it does not matter that he "knows" the partitions and priors of the others. It is only in interpreting the results that these points become significant.

(c) *Finiteness of the Model.* Unlike the finiteness of Γ assumed in Section 7.2, the finiteness of Ω does involve some loss of generality. For example, in describing uncertainty about other people's uncertainty, it is perhaps more natural to allow a continuum of states. Models that do this in full generality and entirely explicitly are, in fact, available; e.g., Mertens and Zamir (1985).

We nevertheless chose the finite model, because the point we make is basically very simple, and has nothing to do with whether the model is finite or infinite. This simplicity might be obscured by an infinite model, with its relatively heavy apparatus of σ-fields, measures, integrals, Radon-Nikodym derivatives, and so on. Such an analysis can probably be carried out in a manner analogous to that presented here; but we have not actually done this, and the possibility of unexpected difficulties cannot be excluded.

In the last analysis, the world *is* usually considered finite; in a sense, finite models do appear more natural. An infinite model is appropriate when there is something to be gained from it, when we are dealing with a phenomenon that is not conveniently expressible in a finite framework. The choice of model is a matter of convenience; here, the finite model is the more convenient and transparent one.

(d) *The Converse.* Let us call the system consisting of Ω, p, the $\mathscr{P}^i$, and $\mathbf{s}$ an *information system.* The main theorem asserts that under Bayesian rationality, every information system corresponds to a correlated equilibrium. The converse also holds: Under Bayesian rationality, every correlated equilibrium corresponds to some information system. More precisely, for each game G and each correlated equilibrium f of G, there is an information system for which it is Bayes rational for the players to play in accordance with $\mathbf{s}$, and the resulting distribution is the same as that of f.

For the proof, note that f itself provides an example of such an information system. We need only define $(\Omega, p) := \Gamma$, $\mathbf{s} := f$, and $\mathscr{P}^i$ to be the partition of Ω generated by f^i (i.e., ω and ω' are in the same element of $\mathscr{P}^i$ iff $f^i(\omega) = f^i(\omega')$).[9] Bayesian rationality is then a restatement of (2.2). Thus under Bayesian rationality, the set of all information systems corresponds precisely to the set of all correlated equilibria.

---

[9]Note that the definitions of Ω, p, and $\mathbf{s}$ go in exactly the opposite direction from the corresponding ones in the proof of the main theorem, where we defined $\Gamma := (\Omega, p)$, $f := \mathbf{s}$. But the proof of the main theorem is not simply a mirror image of this one, since there p^l was not necessarily generated by f^l.

(e) *Exogeneity and Endogeneity.* As we indicated at the beginning of the previous section, it is probably best to think of the information system as exogenous. But a case could be made for another view. While a part of each player's information is undoubtedly generated by overt signals from the outside world, another part is obtained by reasoning about how other players reason. For example, this is so for the information that a strictly dominated strategy will not be used. Some readers might prefer to call that kind of information "endogenous." In actuality, much information is a mixture of both kinds, which it is not easy to untangle.

The distinction between "exogenous" and "endogenous" is often useful in economics, but it should not be pushed too far. In the natural sciences, such a distinction is little used.[10] When discussing the motion of the planets, should the motion of Uranus be considered "exogenous" and that of Neptune "endogenous", or the other way around? Perhaps it is not so terribly important. What *is* important is the relationship between the motions.

The purpose of this paper is to record the observation that common knowledge of rationality implies correlated equilibrium. For this, it is not very important whether the information of the players — or even their actions — are considered exogenous or endogenous.

(f) *Mixed Strategies.* In view of the fact that the players may wish to use mixed strategies, the reader might question the assumption that each player knows his own action. There is no problem if, as usual, one thinks of a mixed strategy simply as a random device for helping the player make up his mind, so that in the end he does know which action he takes. But it is conceivable that a player can actually commit himself to a mixed strategy before he knows its pure outcome; in that case, the assumption in question seems formally incorrect.

The simplest way to circumvent the difficulty is to say that if indeed it is possible for a player i to commit himself irrevocably to

---

[10]We are aware of the pitfalls of blindly applying the methodology of one science to another. But neither should one go to the other extreme, of blindly rejecting all parallels.

a definite mixed strategy, then such a commitment should itself be considered an action — a member of S^i — and then, of course, there is no difficulty. It should be noted that this will not change the set of possible outcomes; adding an action whose payoffs are a convex combination of the payoffs to other actions does not change the set of correlated equilibria.

(g) *Alternative Definitions of Correlated Equilibrium.* In our 1974 definition of correlated equilibrium, each player i is endowed with an information partition $\mathscr{I}_i$ on the underlying probability space (the space of possible outcomes of the randomizing device). After finding out which element of $\mathscr{I}_i$ took place, i may take any action he wishes. Thus an equilibrium is a function from outcomes of the randomizing device to action n-tuples, whose ith component is $\mathscr{I}_i$-measurable (each player knows which action he takes). The equilibrium condition is as in Section 7.2 above.

At first blush, this definition appears different from that of this paper (2.1), since here a player is told *only* what action he takes, and is given no further information. For fixed $\mathscr{I}_i$, the definitions are indeed different. But if the $\mathscr{I}_i$ are allowed to vary, the set of correlated equilibrium distributions obtained under the 1974 definition is the same as that of this paper.

To see this, note that the 1974 set-up is formally identical with that of Section 7.3 above, with $\mathscr{I}_i$ instead of $\mathscr{P}^i$. Thus the fact that every c.e.d. in the 1974 sense is also a c.e.d. in the current sense is simply a restatement of our main theorem. The converse is immediate, since we may, if we wish, take $\mathscr{I}_i$ to consist of the coarsest partition of the underlying probability space for which i's action is measurable.

The reason that in this paper we chose the definition without the $\mathscr{I}_i$ is that it makes the explicit calculation of the set of all c.e.d.'s more transparent.

(h) *Common Knowledge and Universality.* The formulation of our result in the introduction assumes common knowledge (c.k.) of rationality. The formal treatment of Section 7.3 assumes *universal* rationality, i.e., rationality at each point of the state space Ω. These two formulations are equivalent. Indeed, our 1976 treatment of c.k. implies that the universal event Ω is itself c.k. (i.e., anything that is true at all possible states of the world is common knowledge). Conversely, a c.k. event can without loss of generality be assumed universal, because one can always restrict the universe to the smallest c.k. event (i.e., to that member of the meet $\wedge \mathscr{P}^i$ of the information partitions that contains the true state ω of the world).

7.5. The Common Prior Assumption

This section is devoted to a discussion of the only element of our model that is not a tautological consequence of the Bayesian approach: the common prior assumption (CPA for short; see Section 7.3).

Common priors are explicit or implicit in the vast majority of the differential information literature in economics and game theory. As soon as one writes, "let p (rather than p^i) be the probability of...", one has assumed common priors. The assumption is pervasive in the enormous literature on rational expectations, trading in securities, bargaining under incomplete information, auctions, repeated games, signalling, discrimination, insurance, principal-agent, moral hazard, search, entry and exit, bankruptcy, what have you. Citing the relevant papers would make our references longer than our text. Occasionally the definitions do pay lip-service to the possibility of distinct priors p^i; but usually this is quickly abandoned, and in the theorems and examples, one returns to common priors.[11]

In game theory, the standard representation of extensive games[12] involves moves of "chance" or "nature" where each alternative has a single probability, common to all players. The standard formulation of Nash equilibrium (1951) uses mixed strategies for which the probabilities are common to all players. Both these definitions have been used hundreds — perhaps thousands — of times in economic applications, almost without question.

[11] A notable exception is Harrison and Kreps (1978).
[12] von Neumann and Morgenstern (1944), Kuhn (1953).

Why is this so? Why has the economic community been unwilling, in practice, to accept and actually use the idea of truly *personal* probabilities, in much the same way that it did accept the idea of personal utility functions? After all, in the development of Savage (1954), both the utilities and the probabilities are derived separately for each decision maker. Why were utilities accepted as personal, but probabilities not?

Perhaps the most basic reason is that utilities directly express tastes, which are inherently personal. It would be silly to talk about "impersonal tastes", tastes that are "objective" or "unbiased." But it is not at all silly to talk about unbiased probability estimates, and even to strive to achieve them. On the contrary, people are often criticized for wishful thinking — for letting their preferences color their judgment. One cannot sensibly ask for expert advice on what one's tastes should be; but one may well ask for expert advice on probabilities.[13]

[13]That Bayesian decision theory a la Savage derives both utilities and probabilities from preferences does not imply that it does not discriminate conceptually between these two concepts. Savage used preferences to get a rigorous handle on probabilities, a coherent formulation free of inner contradictions. At the same time, he strove, to the extent possible, to rid his probabilities of personal elements, to make them fit for use not only in decision theory but also for scientific induction. He was fond of pointing out that after sufficiently many observations, all "reasonable" priors lead to similar conclusions. His postulate P4 (roughly speaking, that the probability of an event is independent of the prize offered contingent on that event) can only be understood in terms of a probability concept that has an existence of its own in the decision maker's mind, quite apart from preferences on acts. He wrote that "...the...view sponsored here does not leave room for optimism or pessimism...to play any role in the person's judgment" (1954, p. 68). And again, "All views of probability are rather intimately connected with one another. For example, any necessary view can be regarded as an extreme personalistic view in which so many criteria of consistency have been invoked that there is no role left for individual judgment" (1954, p. 60).

Savage is dead — so much the worse for us — and one can only speculate as to how he would have regarded the CPA. As far as we know, he never dealt formally with more than one decision maker at a time. The CPA can be brought formally into his framework via an axiom that says that under common knowledge of rationality, risk averse agents with precisely the same information will never bet against each other (one must, in some appropriate way, exclude hedging).

On a more pragmatic level, the CPA expresses the view that probabilities should be based on information; that people with different information may legitimately entertain different probabilities, but there is no rational basis for people who have always been fed precisely the same information to do so.

Differently put, a major message of modern economic theory is the vital importance of information in economic activity, and of differences in information. Under the CPA, differences in probabilities express differences in information *only*. Thus the CPA enables one to zero in on purely informational issues in analyzing economic (and other interactive) models with uncertainty.

If, in spite of all this, we abandon the CPA, all is not lost. One can define a *subjective correlated equilibrium* as in Section 7.2, replacing the single probability p on Γ by n different probabilities $p^1, \ldots, p^n$. In the main theorem, then, one must only replace "correlated equilibrium distribution" by "subjective correlated equilibrium distribution."

While such an approach is mathematically perfectly consistent, it yields results that are far less sharp than those obtained with common priors. For example, if a 2-person 0-sum game G in strategic form has a unique pair s of optimal (maxmin and minmax) strategies, which are moreover pure, then one would certainly expect rational players to play s; the reasoning leading to this conclusion, which was spelled out in detail by von Neumann and Morgenstern (1944), is extremely compelling. And indeed, s is the only correlated equilibrium point[14] in g. But, it is far from the only subjective correlated equilibrium point. For example, in the game of Figure 7.6, the unique pair of optimal strategies is (top, left); it leads to the

---

Contrary to the modern vogue, Savage was not a minimalist; he did not try to make his axioms as few and weak as possible, but as useful as possible. At one point he wrote that "the personalistic view incorporates all... criteria for reasonableness in judgment known to me, and ... when any criteria that may have been overlooked are brought forward, they will be welcomed ..." (1954, p. 67). It's just possible that he would have welcomed the CPA.

[14]More generally, in any 2-person 0-sum game, all correlated equilibria are convex combinations of pairs of optimal strategies.

1	2	2
−2	9	−9
−2	−9	9

Figure 7.6

0, 0	0, 0	0, 0
0, 0	$\frac{1}{3}, \frac{1}{6}$	$\frac{1}{6}, \frac{1}{3}$
0, 0	$\frac{1}{6}, \frac{1}{3}$	$\frac{1}{3}, \frac{1}{6}$

Figure 7.7

payoff +1 for the row player, and −1 for the column player. But the subjective distribution indicated in Figure 7.7 is a subjective correlated equilibrium; its payoff is +3 for *each* player (in spite of the game being zero sum!), and it makes no use at all of the unique optimal strategies.

This game is not a fluke. Quite generally, the concept of subjective correlated equilibrium places very few restrictions on the possible outcomes. To get the flavor of this, note that for *any* two-person game, not necessarily zero-sum, in which there are no weakly dominated actions, there is a subjective correlated equilibrium at which both players assign positive probability to each outcome (action pair); and there is also a subjective correlated equilibrium that gives each of the two players an expected payoff as close as you wish to his maximum possible payoff. Such results indicate that the subjective correlated equilibrium is a relatively "weak" concept, giving little information; and that while logically consistent, it involves a conceptual inconsistency between the players, which distorts and hides the conflict of interests that is the subject of game-theoretic analysis.

The case of distinct priors was considered by John Harsanyi (1967) in his path-breaking papers on games of incomplete information; he called it the "inconsistent case." The idea of subjective randomization was introduced and developed in our 1974 paper; this idea depends essentially on distinct priors. Neither idea had any considerable echo. Apparently, economists feel that this kind of analysis is too inconclusive for practical use, and side-steps the major economic issues.

7.6. Randomness as an Expression of Ignorance

In the traditional view of strategy randomization, the players use a randomizing device, such as a coin flip, to decide on their actions. This view has always had difficulties. Practically speaking, the idea that serious people would base important decisions on the flip of a coin is difficult to accept. Conceptually, too, there are problems. The reason a player must randomize in equilibrium is only to keep others from deviating; for himself, randomizing is unnecessary.

The first to break away from the idea of explicit randomization was J. Harsanyi (1973). He showed that if the payoffs to each player i in a game are subjected to small independent random perturbations, known to i but not to the other players, then the resulting game of incomplete information has *pure* strategy equilibria that correspond to the mixed strategy equilibria of the original game. In plain words, nobody really randomizes. The appearance of randomization is due to the payoffs not being exactly known to all; each player, who does know his own payoff exactly, has a unique optimal action against his estimate of what the others will do.

Here we take this reasoning one step further. Even without perturbed payoffs, the players simply do not know what actions the other players will take. In "matching pennies", each player knows very well what he himself will do, but ascribes $\frac{1}{2} - \frac{1}{2}$ probabilities to the other's actions, and knows that the other ascribes those probabilities to his own actions.

With this view, mixed strategy equilibria appear quite special and rather unnatural. They imply that the players always act as if they all had the same beliefs about what all other players will do, and as if these beliefs were common knowledge. (The actions themselves, of course, need not be common knowledge, except in the case of pure

.3	.3
.3	.1

Figure 7.8

strategy equilibria.) In particular, they act as if[15] each player i always knew exactly what each other player believes about his (i's) actions.

Correlated strategy n-tuples, on the other hand, do allow for varying beliefs about the beliefs of other players. In the distribution of Figure 7.8, for example, Player 1 may ascribe probabilities that are either $\frac{1}{2} - \frac{1}{2}$ or $\frac{3}{4} - \frac{1}{4}$ to Player 2's actions; and 2 can never know what 1 believes about his (2's) actions. Such phenomena are often of the essence in the analysis of real situations, but they are ruled out by the mixed strategy paradigm.

In games with more than two players, correlation may express the fact that what 3, say, thinks that 1 will do may depend on what he thinks 2 will do. This has no connection with any overt or even covert collusion between 1 and 2; they may be acting entirely independently. Thus it may be common knowledge that both 1 and 2 went to business school, or perhaps to the same business school; but 3 may not know what is taught there. In that case 3 would think

[15]The phrase "act as if' means that we ignore irrelevant information, information that does not affect the actions of the players. Technically speaking, the corresponding assertions hold when each player conditions on his own action only. Without this caveat, they are false. For example, consider a three-person game in which each player chooses 0 or 1, and then is paid 0 or 1 according to whether the sum of the chosen numbers is even or odd. It is a Nash equilibrium for each player to play 0 or 1 with $\frac{1}{2} - \frac{1}{2}$ probabilities. One explicit scenario for this is that each player decides on his action in accordance with a completely private coin toss. A second scenario is like the first, except that 2 observes l's toss. In that case, 2 believes with certainty that 1 will do what he in fact will do, but this belief is not shared by 3, who still ascribes probabilities $\frac{1}{2} - \frac{1}{2}$ to l's actions. A third scenario is like the second, except that 2's observation of l's toss is transmitted through a noisy channel; in that case, 1 will not know what 2 believes about l's action. (In each case, the scenario is assumed common knowledge.) The point is that for practical purposes, all three scenarios are equivalent, since the additional information never causes any player to change his action.

0, 0, 3	0, 0, 0
1, 0, 0	0, 0, 0

2, 2, 2	0, 0, 0
0, 0, 0	2, 2, 2

0, 0, 0	0, 0, 0
0, 1, 0	0, 0, 3

Figure 7.9

it is quite likely that they would take similar actions, without being able to guess what those actions might be.

For example, consider the game of Figure 7.9 (Aumann, 1974). Players 1, 2, and 3 pick the row, column, and matrix respectively. No mixed strategy equilibria yield any player more than 1. But there is a correlated equilibrium yielding (2, 2, 2): Player 3 picks the middle matrix with certainty, while 1 and 2 pick (top, left) with probability $\frac{1}{2}$, and (bottom, right) with probability $\frac{1}{2}$. This does *not* imply collusion between 1 and 2; it simply means that it is common knowledge that 3 has no idea what they will do, but does think with certainty that they will either play (top, left) or (bottom, right). This, of course, is an extreme situation; it will perhaps be more common to find situations where in 3's estimate, the actions of 1 and of 2 are neither prefectly correlated nor independent. Again, such situations are plentiful, but cannot be described by means of mixed strategies.

7.7. The Literature

The idea presented here was briefly outlined in Appendix 4 of Aumann (1981). Kadane and Larkey (1982) also expressed the idea that the theory of games should be consistent with Bayesian decision theory. But they ignored the fact that a rational player must take into account how other players reason about him, and concluded that under Bayesian theory, nothing at all could be said about what each player thinks others would do. One must, they said, turn to disciplines such as psychology to gain insight into how rational agents should or would play a game.

More recently, Bernheim (1984) and Pearce (1984) have independently developed the theory of rationalizability. This is related to the notion of subjective correlated equilibrium mentioned in Section 7.5 above; a rationalizable outcome is much like one that occurs with positive probability in a subjective correlated equilibrium (see Tan and Werlang, 1984, and Brandenburger and Dekel, 1987). Nevertheless, there are differences; for example, rationalizability as currently defined does not permit one player to perceive the strategies of other players as correlated (see Section 7.6 above).

References

Aumann, R. J. (1974): "Subjectivity and Correlation in Randomized Strategies," *Journal of Mathematical Economics*, 1, 67–96. [Chapter 4]

——— (1976): "Agreeing to Disagree," *Annals of Statistics*, 4, 1236–1239. [Chapter 6]

——— (1981): "Survey of Repeated Games," in *Essays in Game Theory and Mathematical Economics in Honor of Oskar Morgenstern*. Mannheim, Wien, Zurich: Wissenschaftsverlag, Bibliographisches Institut, 11–42.

Bernheim, B. D. (1984): "Rationalizable Strategic Behavior," *Econometrica*, 52, 1007–1028.

Brandenburger, A., and E. Dekel (1987): "Rationalizability and Correlated Equilibria," *Econometrica*, 55, 1391–1402.

Harrison, J. M., and D. M. Kreps (1978): "Speculative Investor Behavior in a Stock Market with Heterogeneous Expectations," *Quarterly Journal of Economics*, 92, 323–336.

Harsanyi, J. C. (1967): "Games with Incomplete Information Played by Bayesian Players," Parts I, II, III, *Management Science*, 14, 159–182, 320–334, 486–502.

Harsanyi, J. C. (1973): "Games with Randomly Disturbed Payoffs: A New Rationale for Mixed Strategy Equilibrium Points," *International Journal of Game Theory*, 2, 1–23.

——— (1977): *Rational Behavior and Bargaining Equilibrium in Games and Social Situations*. Cambridge: Cambridge University Press.

Kadane, J. B., and P. D. Larkey (1982): "Subjective Probability and the Theory of Games," *Management Science*, 28, 113–120.

Kuhn, H. W. (1953): "Extensive Games and the Problem of Information," in *Contributions to the Theory of Games II*, ed. by H. W. Kuhn and A. W. Tucker. Princeton: Princeton University Press, 193–216.

Kalai, E., and D. Samet (1984): "Persistent Equilibria in Strategic Games," *International Journal of Game Theory*, 13, 129–144.

Kohlberg, E., and J. F. Mertens (1986): "On the Strategic Stability of Equilibria," *Econometrica*, 54, 1003–1037.

Kreps, D., and R. Wilson (1982): "Sequential Equilibria," *Econometrica*, 50, 863–894.

Lewis, D. K. (1969): *Convention*. Cambridge: Harvard University Press.

Mertens, J. F., and S. Zamir (1985): "Formulation of Bayesian Analysis for Games with Incomplete Information," *International Journal of Game Theory*, 14, 1–29.

Moulin, H., and J. P. Vial (1978): "Strategically Zero-Sum Games: The Class of Games Whose Completely Mixed Equilibria Cannot be Improved Upon," *International Journal of Game Theory*, 7, 201–221.

Myerson, R. B. (1978): "Refinement of the Nash Equilibrium Concept," *International Journal of Game Theory*, 7, 73–80.

Nash, J. F. (1951): "Non-Cooperative Games," *Annals of Mathematics*, 54, 286–295.

von Neumann, J., and O. Morgenstern (1944): *Theory of Games and Economic Behavior*. Princeton: Princeton University Press.

Pearce, D. G. (1984): "Rationalizable Strategic Behavior and the Problem of Perfection," *Econometrica*, 52, 1029–1050.

Savage, L. J. (1954): *The Foundations of Statistics*. New York: Wiley.

Selten, R. (1975): "Re-examination of the Perfectness Concept for Equilibrium Points in Extensive Games," *International Journal of Game Theory*, 4, 25–55.

Tan, T. C.-C., and S. R. da Costa Werlang (1984): "The Bayesian Foundations of Rationalizable Strategic Behavior and Nash Equilibrium Behavior," unpublished preprint, University of Chicago.

Chapter 8

Long-Term Competition — A Game-Theoretic Analysis*

8.1. Introduction

There have been continuing expressions of interest from a variety of quarters in the development of techniques for modelling national behavior in a long-term context of continuing international rivalry — for short, "long term competition." The most characteristic feature of these models is that they extend over time in a fairly regular or repetitive manner. The underlying structure of possible actions and consequences remains the same, though parameters may vary and balances shift, and the decisions and policies of the national decision makers are by no means constrained to be constant or smoothly-varying, or even "rational" in any precisely identifiable sense. The use of game theory or an extension thereof is obviously indicated, and considerable theoretical progress has been made in this area. But the ability of the theory to handle real applications is still far from satisfactory. The trouble lies less with the *descriptive modelling*, i.e., formulating the "rules of the game" in a dynamic setting, than with the choice of a *solution concept* that will do dynamic justice to the interplay of motivations of the actors. (Game theoreticians, like mathematical economists, have always been more comfortable with static than dynamic models.) Since any predictions,

---

*Co-authored with L.S. Shapley. Reprinted from *Essays in Game Theory in Honor of Michael Maschler* (1994), edited by N. Megiddo, 1–15, with permission from Springer.

recommendations, etc. that a mathematical analysis can produce will likely be very sensitive to the rationale of the solution that is used, and since the big difficulties are conceptual rather than technical, it seems both possible and worthwhile to discuss salient features of the theory without recourse to heavy mathematical apparatus or overly formal arguments, and thereby perhaps make the issues involved accessible to at least some of the potential customers for the practical analyses that we wish we could carry out in a more satisfactory and convincing manner.

Two general types of "solution concept" are distinguished in game theory: *cooperative* notions, such as the core, bargaining set, von Neumann–Morgenstern stable sets, and Shapley value; and *noncooperative* notions, principally the Nash equilibrium point and its variants and elaborations, but including also the max-min solution based on "safety level" or "worst case" considerations. Cooperative notions are appropriate for situations where contracts among players are customarily adhered to and can be made legally binding; noncooperative notions where there is mistrust and no external enforcement mechanisms are available. The long-term international scene is most naturally classified as noncooperative, since there is no effective international jurisdiction in most cases, even in the short run. Adherence to major international agreements is essentially a matter of national self-interest, and to be effective in the long run such agreements must be written to be self-enforcing, i.e., so that it is to the continuing advantage of all sides to adhere to them.

Quite a bit is known about Nash noncooperative equilibria in "continuingly competitive" situations, and we shall review some of this material here. It turns out that individual self-interest in such situations can in fact dictate a kind of cooperative behavior, in many cases, sustained by the fear of "punishment" by the other players for failing to "cooperate" with the general plan — this in spite of the fact that the players have no way of legally binding themselves to carry out such punishment. The ability of the noncooperative theory to describe such arrangements and to account for their stability in a "selfish" world is an encouraging point in its favor. The price that is paid, however, is the high degree of nonuniqueness in the Nash solutions (as revealed in the two theorems described below), which removes from this theory most of its predictive power.

8.2. Repeated Game Models

In this section we shall review some of the known theory of a special kind of "continuingly competitive" game: that of *repeated games.* Given a finite game G in strategic form,[1] we consider an infinite game G^*, each play of which consists of an infinite repetition of plays of G. In each play of G, or "round of G^*" the players are assumed to know the outcomes of all previous rounds. The payoff for G^* may be assumed to be of the limiting average form[2]:

$$\lim_{m\to\infty} \frac{1}{m} \sum_{t=1}^{m} h_t. \tag{2.1}$$

Here h_j is the payoff for G in the j'th round of G^*. Many authors call G^* the *supergame* of G.

An alternative form of the payoff for G^* involves discounting of future payoffs at a positive discount rate:

$$\sum_{t=1}^{\infty} (1-\rho)^t h_t. \tag{2.2}$$

For the time being we shall confine our discussion to the limiting average form (2.1), which treats the future as no less important than the present. Indeed, cumulatively, the future is all important in (2.1), since the contribution from any period of finite length will wash out in the long run. Nothing you actually do makes any difference; only your policies for the indefinite future have any significance. Despite this peculiarity, however, it should be remarked that for many purposes, both technical and conceptual, the limiting-average case behaves like the limit of the discounted-sum case, as the discount rate ρ in (2.2)

[1] a.k.a. "normal form."

[2] A technical difficulty is that this limit need not always exist; this technical difficulty has a technical solution, which we do not wish to get involved with at present.

goes to zero. Thus, used with care, the limiting average form can serve as an approximation to situations where a very low discount rate is appropriate. Long-term competition, almost by definition, would appear to fall into this category.

The basic theorem[3] about supergames states that a *necessary and sufficient condition for* $h = (h^1, \ldots, h^n)$ *to be the payoff vector of some Nash equilibrium point of* G^* *is that it be feasible and individually rational in* G.

Let us explain the key terms in this theorem. A "payoff vector" is simply an n-tuple of real numbers, where n is the number of players. The term "payoff vector" is used because the n coordinates signify the payoffs to the n players. By "feasible" we here mean "feasible in correlated strategies"; that is, a payoff vector is feasible if and only if it is in the convex hull of the set of payoff vectors that can be obtained by having the players play pure strategies. A payoff vector is called "individually rational" if each player receives at least his *min-max payoff*, which is the level of payoff below which he cannot be forced by the remaining players.[4] Finally, a "Nash equilibrium point," or "EP", is an n-tuple of strategies — one for each player in the game — such that each player's strategy is a best response to the $(n-1)$-tuple of the other players' strategies. In other words, no player can improve his own payoff by "defecting" to another strategy while the other players are held fixed.

To clarify the meaning of this theorem, let us see what it says about the well-known "Prisoner's Dilemma." This is the two-player game whose strategic form is of the type given in the following table:

	Player II	
Player I	4, 4	0, 5
	5, 0	1, 1

The set of all feasible payoff vectors is indicated by the horizontally hatched region in Figure 8.1, which is the convex hull of the four

[3]This is a "folk-theorem"; it has never been published, but is well known to most workers in the field.

[4]When there are just two players, min-max = max-min.

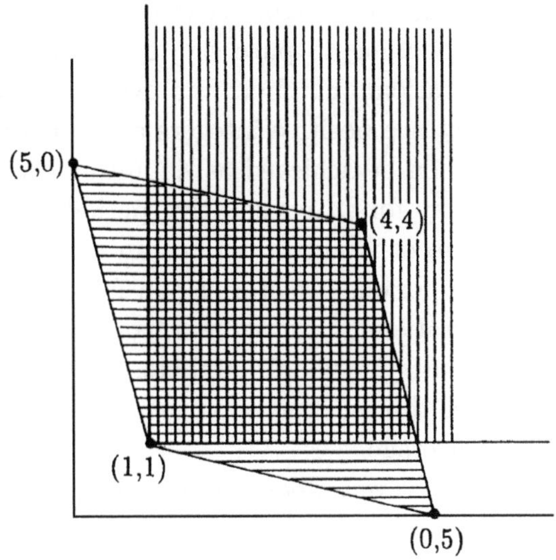

Figure 8.1: Feasible and individually rational payoffs in the Prisoner's Dilemma.

payoffs in the table. Since the minmax payoff to each player is 1, the set of all individually rational payoff vectors is indicated by the vertically hatched region. By the theorem, then, the set of payoff vectors arising from equilibrium points in the supergame is given by the *cross-hatched* region. Note in particular that the point $(4, 4)$ — the traditional "cooperative" outcome — appears as the payoff to an equilibrium point in the supergame of the Prisoner's Dilemma. The proof of the theorem is not difficult, and as the idea of the proof is important to a proper understanding of the situation we shall take a little space to outline it here. The "necessity" part is easily established; it is intuitively clear that equilibrium is not possible if any player is below his guaranteed minimum. The more interesting and significant part of the proof is the "sufficiency."

Assume for simplicity that $n = 2$ (there are just two players). Suppose h is a feasible, individually rational payoff vector. Here we may write

$$h = \sum_{m=1}^{k} \alpha_m h_m,$$

where the α are nonnegative weights that sum to 1 and the h_m are payoff vectors corresponding to pure strategy pairs in G. Suppose first that the α_m are rational numbers and express them in the form $\alpha_m = p_m/q$, where p_m are positive integers and q is their sum. The payoff vector h can then be achieved as a limiting average in G^* by having the players play for p_1 consecutive periods an n-tuple that achieves h_1, then for p_2 consecutive periods an n-tuple that achieves h_2, and so on; after q periods, we start again from the beginning.

If the α_m are irrational, the same effect can be attained by approximating to them by rational numbers, with increasing values of q, and playing once through each approximation in turn, to yield the desired limiting average.

This procedure, however, does not yet describe a Nash equilibrium point in G^*, and in fact does not even describe a pair of supergame strategies. It only describes a particular, feasible course of play. A supergame *strategy* must describe each player's responses to all possible actions of the other player, not only when he "plays along" with a prescribed course of play, such as the one described above, but also when he "defects." This is where the requirement that h be individually rational comes in.

Since h is individually rational, we have for each player i

$$h^i \geqq \max_{\sigma} \min_{\tau} H^i(\sigma, \tau),$$

where H^i is the payoff function to player i in the game G, σ ranges over all mixed strategies of player i, and τ ranges over all mixed strategies of the other player, j. By von Neumann's minimax theorem, there is a mixed strategy $\bar{\tau}$ of j such that for all mixed strategies σ' of i,

$$\min \max H^i(\sigma, \tau) \geqq H^i(\sigma', \bar{\tau});$$

hence in particular

$$h^i \geqq H^i(\sigma', \bar{\tau})$$

for all mixed strategies σ' of i. That means that by playing $\bar{\tau}$, j can hold i down to his max-min value, and *a fortiori* to h^i.

We may now describe an EP in G^* as follows: The players start by playing to obtain an average payoff of h, as outlined above. If at any stage a player i "defects" — i.e., does not play the prescribed choice in G for that round — then starting from the next round, the other player j plays the mixed strategy $\bar{\tau}$ forever after. This will hold i's limiting average payoff down to at most h^i, so that he will have gained nothing by his defection. Thus, h is indeed the payoff to an EP.

8.3. Perfect Equilibrium Points

The above line of proof has been subjected to the following criticism: Though there is no question that the strategy pair as described constitutes an equilibrium point, it is not clear under what circumstances it would ever be used. In particular, it is possible that the strategy $\bar{\tau}$, while holding player i down to his minmax payoff, may also be very[5] disadvantageous to the player using it (or to one of the set of players participating in it, if $n > 2$). The equilibrium point dictates that $\bar{\tau}$ will continue to be played "forever," even if i defects only once. As we pointed out at the end of the previous section, $\bar{\tau}$ is supposed to play the role of a deterrent. But an infinite unremitting repetition of $\bar{\tau}$ seems like an unreasonable response to a single act of defection, except for the fact that — in view of (1) — any finite period of "punishment" is no punishment at all. But by the same token, a single defection is also insignificant in the limit. Thus, the threatened response may still seem unreasonable, especially when, as is often the case, it is disadvantageous or costly to the user, and hence such an unremitting repetition may not be believable as a deterrent. To have a word for these EPs, let us call them "grim."

Let us try to pinpoint the dissatisfaction with grim EPs in a slightly more general framework. The "knowledge" that j (or, more generally, $N\setminus\{i\}$) will respond to a defection on the part of i by an unrelenting stream of $\bar{\tau}$ is what keeps i from defecting; but if i does

[5]Compare [1].

in fact defect, it may no longer be profitable for j (or $N\backslash\{i\}$) to respond with $\bar{\tau}$. This is what makes $\bar{\tau}$ unbelievable.

This kind of reasoning motivated a specialization of the notion of equilibrium point, first considered by R. Selten [2] and called by him a perfect equilibrium point.[6] To define this notion, we must recall more precisely the definition of a "strategy" for the player i in the supergame G^*. This is a function that tells i which pure G-strategy to choose on each round, as a function of what all the players, including i himself, did on all previous rounds. For each positive integer k, define G_k^* to be the "subgame" starting from the k'th period, i.e., after $k-1$ rounds have been played, and continuing indefinitely from that point. Thus, $G^* = G_1^*$. Each n-tuple of strategies in G^*, together with a series of actual actions on the part of all players in the first $k-1$ rounds, induces an n-tuple of strategies in G_k^*. An n-tuple $\sigma = (\sigma_1, \ldots, \sigma_n)$ of strategies in G^* is called a *perfect equilibrium point* (or PEP) if for each k and for each series of actions of the players in the first $k-1$ periods of G^*, the induced n-tuple is an EP of G_k^*.

If we set $k = 1$ we see that a PEP is in particular an EP.

It's easy to see that a grim EP is in general *not* perfect, since if player i defects on round $k-1$ it will in general not be a best response in G_k^* for the other players to "punish" him; it may even be individually irrational. It thus appears that the notion of perfect equilibrium point might hold an answer to the problem of the believability of deterrents. In the next three sections we shall explore this matter somewhat further. To some extent our hopes turn out to be in vain: we shall find in the next section that the payoffs associated with PEPs in G^* are the same as those for grim EPs, though the grim EPs themselves are excluded. The methods may be different, so to speak, and possibly more "believable," but the upshot is the same: there is no narrowing of the class of outcomes that can be sustained in equilibrium.

However, when we modify the payoff in G^* by introducing a positive discount rate (2), which we do in Section 8.5, we find that

---

[6] See also [3].

requiring "perfection" can significantly reduce the set of equilibrium outcomes. Moreover, the concept of believability does appear to play a significant role in the description of the perfect equilibria. Thus, the notion of perfection of equilibria, though not a panacea, does appear to give us a somewhat better handle on some of the problems that we wish to model.

8.4. Characterization of PEPs

This section is devoted to the following theorem:

Theorem 4.1. *The set of payoffs to perfect EPs in G^* coincides with the set of payoffs to ordinary EPs — i.e., it is the set of all feasible, individually rational payoffs in G.*

Again, to gain a good understanding of this theorem it is essential to outline the proof. As before, it is "sufficiency" that is the interesting part of the proof; the "necessity" follows from the previous theorem. We shall find the argument considerably more intricate than before.

To simplify the presentation we again assume that there are only two players. Moreover, in order to make the use of mixed strategies unnecessary, we shall assume that G is not in strategic form, with simultaneous choices by the two players, but is a game of perfect information with a single move for each player and no chance moves. Player I moves first, II is informed of I's move, and then II moves.[7] None of these assumptions are really required for the truth of the theorem, but they do simplify the proof.

Suppose h is a feasible, individually rational payoff vector of G. We shall describe a PEP with payoff h. As in Section 8.2, the description will be couched in terms of a tentative "agreement" on a prescribed course of play. The agreement starts out as before with a sequence of choices which, when adhered to by both players, will lead to the desired limiting average payoff h. Let us call this

[7]Note that the apparent asymmetry of the players disappears in the supergame: in G^* the players move alternately, each with perfect information.

"the cooperative sequence." Next, we shall specify how the players react to a defection — i.e., a departure from the cooperative sequence by one of the players. In the previous proof, the reaction was unrelenting punishment. Here, instead, the PEP strategies will specify that a defection on the part of either player be punished by a sequence of choices by the other that forces the defector's average payoff down to within ε of his max-min value, where ε is a small number that may depend on the "date" of the defection. After the defector has thus been "beaten to within an inch of his life," the punisher relents and prescribed play returns to the cooperative sequence at the point of defection.

If should be noted, however, that not only are defections from the cooperative sequence punished, but also defections from any punishing sequence (in the subgame resulting from an earlier defection) are punished. A player who "should" punish and does not do so will himself be punished. This is what provides the motivation for the punisher actually to carry out the punishment, and so keeps the EP perfect.

The situation is a little complex; in order to convince ourselves that we have actually described a PEP we shall now give a more formal treatment. Without loss of generality we may assume that the number of choices available to I on each move is the same as the number of choices available to II; call the number m. Thus, $M = \{1, \ldots, m\}$ is the set of possible choices of the players at each move. When it's Player I's turn in the n-th round of G^* he has before him the full history of previous moves; this takes the form of a sequence $(x_1, \ldots, x_{2n-2})$, where $x_i \in M$ represents the choice made on the i'th move in G^* (it is I's or II's choice according as i is odd or even). Similarly when II must move, he has before him a sequence $(x_1, \ldots, x_{2n-1})$, where again each $x_i \in M$. Let us call any finite sequence of members of M a *history*. A *strategy* for Player I [Player II] may be defined as a function from histories of even [odd] length to M. Thus, a pair of strategies is simply a function f from the set of *all* histories to M.

Now let $h = (h^I, h^{II})$ be the given feasible individually rational payoff vector. Let $(c_1, c_2, c_3, \ldots)$ be a fixed *cooperative sequence*, i.e., a sequence of moves leading to the payoff h in G^*. Let $\bar{p}$ be a G-strategy for I (i.e., a member of M) that holds II to his max-min payoff in G, and let $\bar{q}(\cdot)$ be a G-strategy for II (i.e., a function from M to M) that holds I to his max-min payoff in G.

We wish to define a strategy-pair f for G^* which is a PEP and whose associated payoff is h. The definition of f will be inductive, based on the length k of the history on which it is being defined. On a history of length 0 we define f to be c_1; this simply means that the PEP will prescribe the choice of c_1 for the first move of Player I. Suppose now that f has been defined on all histories up to length $k-1$; we wish to define it on all histories of length k. Let $(x_1, \ldots, x_k)$ be such a history. We shall say that the ℓ'th move of that history ($1 \leqq \ell \leqq k$) is a *defection* if $x_\ell \neq f(x_1, \ldots, x_{\ell-1})$. If $(x_1, \ldots, x_k)$ contains no defection, we define $f(x_1, \ldots, x_k) = c_{k+1}$. Otherwise, suppose the *most recent* defection in $(x_1, \ldots, x_k)$ occurred at move ℓ. If ℓ and $k+1$ have the same parity — i.e., the player who is about to move is the same as the one who most recently defected — then we define $f(x_1, \ldots, x_k) = c_{k+1}$. If ℓ and $k+1$ have opposite parity, consider first the case in which ℓ is even, i.e., Player II was the last to defect. In that case k is also even, so exactly $k/2$ rounds of G^* have now been completed and it is Player I's turn to move. Consider the average payoff of Player II as measured at the end of each of the rounds $\ell/2+1, \ldots, k/2$, and let $\varepsilon_\ell = 1/\ell$. If any one of these averages is $\leqq \varepsilon_\ell +$ II's max-min value in G, then we define $f(x_1, \ldots, x_k) = c_{k+1}$; otherwise, we define $f(x_1, \ldots, x_k) = \bar{p}$. That means that Player I brings the offending Player II to within ε_ℓ of his max-min payoff and then returns to cooperative play.[8]

Finally, consider the case in which ℓ and $k+1$ have opposite parity and ℓ is odd, i.e., Player I was the last to defect. In that case k is odd, $(k-1)/2$ rounds of G have been completed, Player I has already made his move in the $(k+1)/2$'th round, and it is now II's turn to move. Proceeding as before, we consider I's average payoff as measured at the end of each of the periods $(\ell+1)/2, \ldots, (k-1)/2$. If any of these averages was $\leqq \varepsilon_\ell +$ the max-min value to I in G, then

[8]Note that he never has occasion to look back beyond the most recent defection; the ε_ℓ level of punishment suffices for all past transgressions.

we define $f(x_1, \ldots, x_k) = c_{k+1}$; otherwise, we define $f(x_1, \ldots, x_k) = \bar{q}(x_k)$. As before, that means that II brings the offending Player I to within ε_ℓ of his max-min value, then returns to cooperative play. (The difference is only that, because of the asymmetry in G, II's punishing move must depend on I's move in the same "round.")

This completes the formal description of the PEP that we described informally before; the reader should be able to convince himself that it is in equilibrium, is perfect, and yields the cooperative sequence $(c_1, c_2, \ldots, c_k, \ldots)$ with limiting average payoff h.

8.5. Discounted Payoffs in Repeated Games: Discussion of an Example

Thus far, we have been considering only the limiting average form of payoff for repeated games, corresponding intuitively to a future discount rate of zero. We shall now try to give an idea of how positive discount rates can affect the behavior of EPs and PEPs by studying an apparently simple but surprisingly revealing example.

Consider the following payoff matrix for G in strategic form, the players moving simultaneously:

		Player II	
		L	R
Player I	T	0, 1	$-p$, $-c+1$
	B	1, 0	$-p+1$, $-c$

Here, p and c are positive numbers ("punishment" and "cost"); we may think of them as being rather large. Thus, II may be in a position to damage I severely, but only at a cost to himself that may perhaps be unacceptable.[9]

In the repeated G^*, we shall use the discounted payoffs

$$\sum_{t=1}^{\infty} \alpha^t h_t^I \quad \text{and} \quad \sum_{t=1}^{\infty} \beta^t h_t^{II}$$

[9]One could think of I and II as North Vietnam and the United States in the 1960s.

to Players I and II, respectively, where $0 < \alpha < 1$ and $0 < \beta < 1$. Sometimes we shall further assume that $\alpha \leq \beta$, i.e., that Player I has, if anything, a bigger discount rate (= shorter "utility horizon") than Player II.

As is easily seen, the one-shot game G has a unique EP, namely (B, L), which yields the payoff $(1, 0)$. This means that the strategy-pair in which I always plays B and II always plays L (regardless of history) is a perfect EP of G^*, since obviously no defection, even in a subgame, can ever be profitable.

Player II, however, would naturally prefer the outcome $(0, 1)$, corresponding to (T, R). We shall now investigate under what conditions this outcome can be sustained by an EP, or by a PEP, in the discounted repeated game. Indeed, we shall find that it can be sustained by an EP if and only if $p \geqq 1/\alpha$; and, when $\alpha \geqq \beta$, that it can be sustained by a PEP if and only if $p \geqq 1/\alpha$ and $p/c \geqq (1-\beta)/\alpha\beta$. Thus, whereas the existence of an EP is independent of the cost of the punishment to the punisher (the parameter c), the existence of a PEP is not.

Let us first consider the EP question. We claim that the following "grim" strategy-pair:

$$\begin{cases} \text{I} & \text{plays } T \text{ always} \\ \text{II} & \text{plays } L \text{ so long as I plays } T\text{, but plays } R \text{ forever} \\ & \text{if I ever plays } B \end{cases}$$

is an EP of G^*, provided that $p \geqq 1/\alpha$. Moreover, we claim that if $p < 1/\alpha$ there is no EP sustaining $(1, 1)$. Note first that II will certainly not wish to defect, as he cannot possibly improve on the sequence of payoffs $(1, 1, 1, \ldots)$. On the other hand, if I wishes to improve on his sequence $(0, 0, 0, \ldots)$, his best chance is to deviate to B at some time t_0, then keep playing B forever. This yields him the payoff sequence $(0, \ldots, 0, 1, -p+1, \ldots)$, and this is worth

$$\alpha^{t_0}\left(\frac{1}{1-\alpha} - \frac{p\alpha}{1-\alpha}\right).$$

Since this is profitable to him *if and only if* $1 > p\alpha$ the truth of our claims is now evident.

For a numerical example, let $p = 2$. Then if $\alpha < .5$ there will be no EP sustaining the $(0, 1)$ outcome, as the rewards for defecting will outweigh any possible punishment. But if $\alpha \leq .5$, the strategy pair given above is clearly an EP.

Nothing in this result depends on the values of c or β. Yet, intuitively, one feels that the credibility of II's "threat," with which he extracts such a favorable outcome, ought to be *very* dependent on its cost. Our next object will be to show that a PEP that sustains the $(0, 1)$ outcome is not possible for large values of c.

First let us give an example of such a PEP. It happens that we can define it in a very simple way, making the instructions to the two players almost independent of the history:

1. In the first round, play (T, L).
2. If the choices in round $t-1$ were (T, L) or (B, R), play (T, L) in round t.
3. If the choices in round $t-1$ were (T, R) or (B, L), play (B, R) in round t.

The cooperative sequence resulting from this strategy pair is just a repetition of (T, L); this is worth $\beta/(1-\beta)$ to II and 0 to I. In checking for the PEP property, it is sufficient to look merely at deviations that occur in the first round of a typical subgame G_t^*. Suppose Player I defects when he is supposed to play T. His best possible payoff sequence from then on is $(1, -p+1, -p+1, \ldots)$, which is worth

$$\alpha^t\left(\frac{\alpha}{1-\alpha} - \frac{\alpha^2 p}{1-\alpha}\right)$$

to him. So if $p \geqq 1/\alpha$ he will not have any incentive to defect. Player II likewise will not defect when he is supposed to play L, as he cannot possibly improve on the payoff sequence $(1, 1, 1, \ldots)$. When I is supposed to play B, a defection could yield him at best the sequence $(-p, -p+1, 0, 0, \ldots)$; this is clearly inferior for all $p \geqq 0$ to the prescribed sequence $(-p+1, 0, 0, \ldots)$. Finally, when II is called upon to play R at the beginning of G_t^*, he will have to compare his prescribed payoff sequence $(-c, 1, 1, 1, \ldots)$ with sequences like

$(0, -c, 1, 1, 1, \ldots)$, $(0, 0, -c, 1, 1, 1, \ldots)$, etc., which he can obtain by defecting for 1, 2, etc. rounds, or even the sequence $(0, 0, 0, \ldots)$ which he can obtain by perpetually defecting. In the discounted sum, these "heresies" are worth

$$\beta^k(-c\beta + \beta^2/(1-\beta)), \quad k = t+1, t+2, \ldots,$$

or 0, while "orthodoxy" is worth

$$\beta^t(-c\beta + \beta^2/(1-\beta)).$$

So if $c \leqq \beta(1-\beta)$, II cannot gain by defecting. The given strategy pair is therefore a PEP on the assumptions that $p \geqq 1/\alpha$ and $c \leqq \beta/(1-\beta)$, as diagrammed in Figure 8.2.

To wrap up our example, it is necessary to show that there are significant cases where, because of the positive discount rate, a PEP does *not* exist. Showing nonexistence is a more difficult undertaking, because in general a PEP can be a very complex thing. In particular, while pure G-strategies have sufficed up to now, we cannot ignore the possible use of mixed strategies against a defection. In our example,

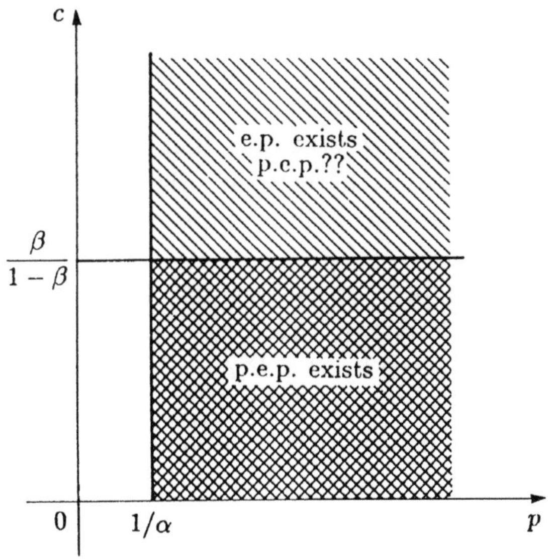

Figure 8.2

if p and c are both large numbers, the threat of a small probability of using R may be enough to keep I in line while holding the (expected) cost to a level that II can accept.[10] However, it would be out of place in this discussion to develop the elaborate technical apparatus of mixed strategies just for the sake of one example, whose purpose is only illustrative. Instead, we shall adopt a far simpler expedient, called "convexification in pure strategies," which is more or less equivalent to the introduction of mixed strategies.

In our example, this convexification merely means allowing Player II the option of "scaling down" his punishment by giving him a continuum of strategies in G, as follows:

		Player II	
		L	$R\lambda$ $(0 < \lambda \leqq 1)$
Player I	T	0, 1	$-\lambda p, -\lambda c + 1$
	B	1, 0	$-\lambda p + 1, -\lambda c$

Here, $\lambda = 1$ corresponds to the old R and $\lambda = 0$ corresponds to the old L. (However, we still indicate the latter choice by a separate column in the matrix.) Playing $R\lambda$ has much the same effect as playing a mixed strategy $\{R$ with probability λ, L with probability $1-\lambda\}$, and it can be shown (though we shall not do it here) that if the new G^* has no PEP in *pure* strategies that sustains the cooperative sequence $((T, L), (T, L), \ldots)$, then the original G^* (with the same values of α, β, p, c) has no PEP in pure or mixed strategies that sustains that sequence.

Consider now a play of the revised G^*, with II making the sequence of moves $\Lambda = (\lambda_1, \lambda_2, \ldots, \lambda_t, \ldots)$. The total punishment received by I is then given by $P = P(\Lambda) = \sum_{t=1}^{\infty} \lambda_t \alpha^t p$, and the total cost incurred by II is $C = C(\Lambda) = \sum_{t=1}^{\infty} \lambda_t \beta^t c$. We now bring in the assumption, not used until now, that $\alpha \leqq \beta$. This implies that

[10] This is a realistic consideration for the world of nuclear politics and arms races, where the pressure of the nuclear deterrent is felt in every situation that creates any perceptible risk that the situation might escalate out of control.

$\sum_{t=1}^{\infty} \lambda_t(\beta^t - \alpha^t) \geqq 0$, so that $P/C \leqq p/c$. This inequality shows that it is most efficient, in terms of the damage/cost ratio, for II to punish *immediately*; he thereby minimizes his cost for a given level of deterrence. It follows that the game has a PEP, of the type described above, whenever there is any number λ such that λp and λc satisfy the inequalities

$$\lambda p \geqq 1/\alpha, \quad \lambda c \leqq \beta/(1-\beta).$$

This is illustrated in Figure 8.3. As we can see, there is a critical ratio of c to p, namely

$$R_0 = \frac{a\beta}{1-\beta},$$

above which no such λ can be found.

For a numerical example, if $\alpha = \beta = .75$ then $R_0 = 2.25$. If $p = 100$ and $c = 200$ then $c/p < R_0$, and we may, for example, choose $\lambda = .014$, giving us $\lambda c = 2.8 < \beta/(1-\beta) = 3$ and $\lambda p = 1.4 >$

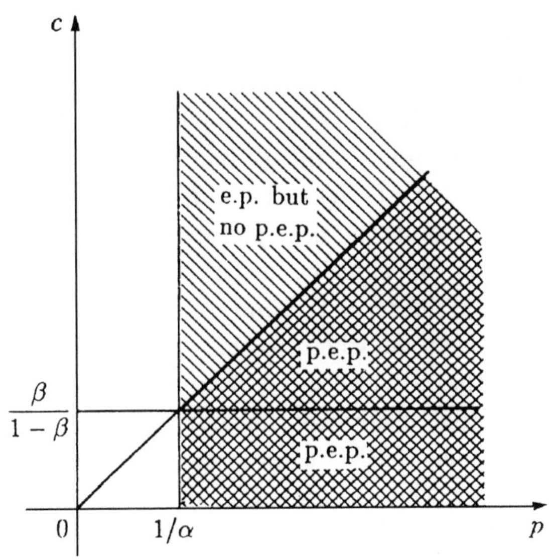

Figure 8.3

$1/\alpha = 1.33$. So a 1.4 percent chance of II using his threat strategy R after a defection by I sustains the perfect equilibrium at $(0, 1)$.

We can also make the converse argument. If (c, p) is *not* in the cross-hatched region indicated in Figure 8.3, then there is no way for II to inflict any given amount P of punishment without incurring a cost of more than $R_0 P$. By the foregoing, it is clear that this is above the cost that he can "afford"; in other words, he would prefer to accept his max-min payoff of 0 forever, rather than carry out the requisite threat. So a PEP cannot exist.

References

1. R. J. Aumann, "Acceptable Points in General Cooperative n-Person Games," in *Contributions to the Theory of Games IV*, Annals of Math. Study 40, Princeton University Press, 1959, 287–324. [Chapter 1]
2. R. Selten, "Spieltheoretische behandlung eines oligopolmodells mit nachfrageträgheit," *Zeitschrift für die gesamte Staatswissenschaft* 12 (1965), 301-324.
3. R. Selten, "Re-examination of the perfectness concept for equilibrium points in extensive games," *International J. Game Theory* 4 (1975), 25–55.

Note added in proof. Professor Jacek Krawczyk of the Victoria University of Wellington has pointed out an error on Page 214. If Player I defects when he is supposed to play T, his best possible payoff sequence from then on is $(1, -p+1, 1, -p+1, 1, \ldots)$, and not as written there. The given strategy pair is therefore a PEP on the assumptions that $p \geq 1+(1/\alpha)$ and $c \leq \beta/(1-\beta)$, and not as written on Page 215. As a result, the neat picture presented in Figure 8.3 is called into question. Specifically, an EP certainly exists in the strip between the vertical lines $p = 1/\alpha$ and $p = 1+(1/\alpha)$, but it is unclear whether a PEP exists anywhere in this strip. The exact placement of the diagonal that divides EP's from PEP's becomes similarly unclear. Nevertheless, it appears that our basic point remains valid: that for the existence of PEP's, the cost-to-punishment ratio is the determining factor, whereas for EP's, only the magnitude of the punishment matters — the cost of the punishment to the punisher is irrelevant.

Chapter 9

Game-Theoretic Analysis of a Bankruptcy Problem from the Talmud*

Dedicated to the memory of Shlomo Aumann, Talmudic scholar and man of the world, killed in action near Khush-e-Dneiba, Lebanon, on the eve of the nineteenth of Sivan, 5742 (June 9, 1982).

For three different bankruptcy problems, the 2000-year-old Babylonian Talmud prescribes solutions that equal precisely the nucleoli of the corresponding coalitional games. A rationale for these solutions that is independent of game theory is given in terms of the Talmudic principle of equal division of the contested amount; this rationale leads to a unique solution for all bankruptcy problems, which always coincides with the nucleolus. Two other rationales for the same rule are suggested, in terms of other Talmudic principles. (Needless to say, the rule in question is not proportional division.) *Journal of Economic Literature Classification Numbers*: 022, 026, 031, 043, 213. © 1985 Academic Press, Inc.

9.1. Introduction

A man dies, leaving debts $d_1, \ldots, d_n$ totalling more than his estate E. How should the estate be divided among the creditors?

*Co-authored with M.B. Mashler. Reprinted from *Journal of Economic Theory* (1985), 36(2), 195–213, with permisson from Elsevier.

A frequent solution in modern law is proportional division. The rationale is that each dollar of debt should be treated in the same way; one looks at dollars rather than people. Yet it is by no means obvious that this is the only equitable or reasonable system. For example, if the estate does not exceed the smallest debt, equal division among the creditors makes good sense. Any amount of debt to one person that goes beyond the entire estate might well be considered irrelevant; you cannot get more than there is.

A fascinating discussion of bankruptcy occurs in the Babylonian Talmud[1] (Kethubot 93a). There are three creditors; the debts are 100, 200, and 300. Three cases are considered, corresponding to estates of 100, 200, and 300. The *Mishna*[2] stipulates the divisions shown in Table 9.1.

The reader is invited to study Table 9.1. When $E = 100$, the estate equals the smallest debt; as pointed out above, equal division then makes good sense. The case $E = 300$ appears based on the different — and inconsistent — principle of proportional division. The figures for $E = 200$ look mysterious; but whatever they may mean, they do not fit any obvious extension of either equal *or* proportional division. A common rationale for all three cases is not apparent.

Table 9.1

		Debt		
		100	200	300
	100	$33\frac{1}{3}$	$33\frac{1}{3}$	$33\frac{1}{3}$
Estate	200	50	75	75
	300	50	100	150

[1]A 2,000-year-old document that forms the basis for Jewish civil, criminal, and religious law.

[2]The basic text that forms the starting point for the discussions recorded in the Talmud.

Over two millenia, this Mishna[3] has spawned a large literature. Many authorities disagree with it outright. Others attribute the figures to special circumstances, not made explicit in the Mishna. A few have attempted direct rationalizations of the figures as such, mostly with little success. One modern scholar, exasperated by his inability to make sense of the text, suggested errors in transcription.[4] In brief, the passage is notoriously difficult.

This paper presents a game-theoretic analysis of the general bankruptcy problem, for arbitrary $d_1, \ldots, d_n$ and E. We obtain (Section 9.6) an explicit characterization of the nucleolus of the coalitional game that is naturally associated with this problem. For the three cases considered in the Talmud, the nucleolus prescribes precisely the numbers of Table 9.1 — those in the Mishna.

Of course, it is unlikely that the sages of the Mishna were familiar with the general notion of a coalitional game, to say nothing of the nucleolus. In Sections 9.3, 9.4, and 9.5, we present three different justifications of the solution to the bankruptcy problem that the nucleolus prescribes, in terms that are independent of each other and of game theory, and that were well within the reach of the sages of the Mishna. The justifications also fit in well with other Talmudic principles, a consideration that is no less significant than innate reasonableness in explaining the text.

To emphasize the independence from game theory, we start with the non-game theoretic analysis. In the research process, however, the order was reversed. Only after realizing that the numbers in the Mishna correspond to the nucleolus did we find independent rationales. Without the game theory, it is unlikely that we would have hit on the analysis presented in Sections 9.2 through 9.5.

---

[3]The word "Mishna" is used both for the entire text on which the Talmud is based, and for specific portions of it dealing with particular issues. Similar ambiguities occur in many languages. One may say "My son studied law" as well as "Yesterday Congress passed a law."

[4]Lewy [8, p. 106], near the end of the long footnote.

This paper addresses two related but distinct questions, "what" and "why." The Mishna explicitly gives only three numerical examples; the first, most basic question is, *what* general rule did it have in mind, what awards would it actually assign to the creditors in an arbitrary bankruptcy problem? The second question is *why* did it choose the rule that it chose, what reasoning guided Rabbi Nathan (the author of this particular Mishna)? We have three different answers to the "why" question, all of them leading to the same answer to the "what" question. And while we cannot be sure that any particular one of our answers to the "why" question really represents Rabbi Nathan's thinking, all of them together leave little doubt that our answer to the "what" question is the correct one.

It is hoped that the research reported here will be of interest in two spheres — in the study of the Talmud, and in game theory. In this paper we concentrate on the mathematical, game-theoretic side. For motivation we will, when appropriate, present and explain underlying Talmudic principles; but there will be no careful textual analysis of Talmudic passages, no lengthy citation of authorities.[5] An analysis of the latter kind — concentrating on the validity of our explanation from the source viewpoint, but skimping on the mathematics — will be published elsewhere.[5a]

9.2. The Contested Garment

A famous Mishna (Baba Metzia 2a) states: "Two hold a garment; one claims it all, the other claims half. Then the one is awarded $\frac{3}{4}$, the other $\frac{1}{4}$."

The principle is clear. The lesser claimant concedes half the garment to the greater one. It is only the remaining half that is

[5]In particular, this article should on no account be used as a source for Talmudic law. In citing Talmudic dicta, we mention only those of their aspects that are directly relevant to the matter at hand; additional conditions and circumstances are omitted. For more information the reader should refer to the sources, which we have been at pains to cite explicitly.

[5a]Aumann, Y., "On the Man with Three Wives," *Moriah* 22, 3–4, Tevet 5759, 1999, pp. 98–107; in Hebrew.

at issue; this remaining half is therefore divided equally.[6] Note that this is quite different from proportional division.

Let us transpose this principle to the 2-creditor bankruptcy problem with estate E and claims d_1, d_2. The amount that each claimant i concedes to the other claimant j is $(E - d_i)_+$, where

$$\theta_+ = \max(\theta, 0).$$

The amount at issue is therefore

$$E - (E - d_1)_+ - (E - d_2)_+;$$

it is shared equally between the two claimants, and, in addition, each claimant receives the amount conceded to her by the other one. Thus the total amount awarded to i is

$$x_i = \frac{E - (E - d_1)_+ - (E - d_2)_+}{2} + (E - d_j)_+. \tag{2.1}$$

We will say that this division (of E for claims d_1, d_2) is prescribed by the CG (contested garment) principle.[7,8]

If one views the solution as a function of E, one obtains the following process. Let $d_1 \leq d_2$. When E is small, it is divided equally. This continues until each claimant has received $d_1/2$. Each additional dollar goes to the greater claimant, until each claimant has received all but $d_1/2$ of her claim. Beyond that, each additional dollar is again divided equally. Note that the principle is *monotonic*, in the sense that for fixed claims d_1 and d_2, each of the two awards is a monotonic function of the estate E.

---

[6] This explanation is explicit in the eleventh-century commentary of Rabbi Shlomo Yitzhaki (Rashi). Alternatively, one could say that the claims total $1\frac{1}{2}$, whereas the worth of the garment is only 1; the loss is shared equally.

[7] An additional instance of this rule may be found in the Tosefta to the first chapter of Baba Metzia, where $E = d_1 = 1$, $d_2 = \frac{1}{3}$. (The Tosefta is a secondary source that is contemporaneous with the Mishna.)

[8] Alternatively, one may argue that neither claimant i can ask for more than $\min(E, d_i)$. If each claimant is awarded this amount, the total payment may exceed the estate; the excess is deducted in equal shares from the claimants' awards. This procedure leads to the same payoff as (2.1).

The legal circumstances of the contested garment are somewhat different from those of bankruptcy. In the garment case, there is uncertainty about the validity of the claims; they cannot both be justified. In the bankruptcy problem, all claims are definitely valid; there simply is not enough money to go around. Therefore, some authorities have held that the Mishna in Baba Metzia (about the garment) is not relevant to the bankruptcy problem.

While this certainly constitutes an important difference between the cases, it is not clear why it would make the principle of equal division of the contested amount inapplicable. Indeed, the early medieval authority Rabbi Hai Gaon (10th century) did express the opinion[9] that the Mishna in Kethubot (about bankruptcy) should be explained on the basis of that in Baba Metzia. He did not, however, make an explicit connection, and in subsequent years, this line of attack was abandoned.

9.3. Consistency

A *bankruptcy problem* is defined as a pair $(E;\, d)$, where $d = (d_1, \ldots, d_n)$, $0 \leq d_1 \leq \cdots \leq d_n$ and $0 \leq E \leq d_1 + \cdots + d_n$. A *solution* to such a problem is an n-tuple $x = (x_1, \ldots, x_n)$ of real numbers with

$$x_1 + \cdots + x_n = E$$

(x_i is the amount assigned to claimant i). A solution is called *CG-consistent*, or simply *consistent*, if for all $i \neq j$, the division of $x_i + x_j$ prescribed by the contested garment principle for claims d_i, d_j is (x_i, x_j).

Intuitively, a solution is consistent if any two claimants i, j use the contested garment principle to divide between them the total amount $x_i + x_j$ awarded to them by the solution. It may be verified that the solutions in Table 9.1 are consistent.

[9]Quoted by Rabbi Isaac Alfasi (1013–1103) in his commentary on our Mishna in Kethubot.

Theorem A. *Each bankruptcy problem has a unique consistent solution.*

Proof. First we prove that there is at most one consistent solution. If there were more, we could find consistent solutions x and y, and creditors i and j, with $y_i > x_i, y_j < x_j$, and $y_i + y_j \geq x_i + x_j$. Consistency implies that if just i and j are involved, the CG principle awards y_j to j when the total estate is $y_i + y_j$, and x_j when it is $x_i + x_j$. Since $y_i + y_j \geq x_i + x_j$, the monotonicity of the CG principle then implies $y_j \geq x_j$, contradicting[10] $y_j < x_j$.

To show that there is at least one consistent solution, we exhibit it as a function of the estate E (for fixed debts $d_1, \ldots, d_n$). Let us think of the estate as gradually growing. When it is small, all n claimants divide it equally. This continues until 1 has received $d_1/2$; for the time being she[11] then stops receiving payments, and each additional dollar is divided equally between the remaining $n - 1$ claimants. This, in turn, continues until 2 has received $d_2/2$, at which point *she* stops receiving payments for the time being, and each additional dollar is divided equally between the remaining $n - 2$ claimants. The process continues until each claimant has received half her claim. This happens when $E = D/2$, where

$$D := d_1 + \cdots + d_n = \text{the total debt.}$$

When $E \geq D/2$, the process is the mirror image of the above. Instead of thinking in terms of i's award x_i, one thinks in terms of her loss $d_i - x_i$, the amount by which her award falls short of her claim. When the total loss $D - E$ is small, it is shared equally between all creditors, so that creditor i receives her claim d_i less $(D - E)/n$. The

[10]This part of the proof was generated at a seminar presentation at the IMA in Minneapolis; it replaces a more devious proof that we previously had. Mainly responsible are Y. Kannai and D. Kleitman.

[11]While we are sympathetic with the feminist movement, the reader should not conclude from our use of "she" that we write half our papers in the feminine gender. Much of the Talmud is couched in terms of case law; and while the passage under discussion does form the basis of bankruptcy law in general, the creditors in this particular case were women.

creditors continue sharing each additional dollar of total loss equally, until 1 has lost $d_1/2$ (which is the same as receiving $d_1/2$). For the time being she then stops losing, and each additional dollar of total loss is divided equally between the remaining $n-1$ claimants. This, in turn, continues until 2 has lost $d_2/2$ (= received $d_2/2$), at which point *she* stops losing for the time being, and each additional dollar is divided equally between the remaining $n-2$ claimants. The process continues until each claimant has lost half her claim, which happens when $E = D/2$. This is precisely to where we got in describing the first part of the procedure; we have dug the tunnel from both its ends, and have met in the middle.

It will be useful to give an alternative description of the procedure, in terms of increasing award. Recall that when E was slightly less than $D/2$, claimant n was receiving all of each additional dollar of the estate. She continues to do so as E passes $D/2$, until she has received a total of $d_n - (d_{n-1}/2)$, i.e., all but $d_{n-1}/2$ of her claim. At this point, $n-1$ reenters the picture, and each additional dollar is shared equally between n and $n-1$. This continues until n and $n-1$ have each received all but $d_{n-2}/2$ of their claims (which happens at the same instant). At this point $n-2$ reenters the picture, and so on. Creditor 1 reenters the picture when *all* creditors have received all but $d_1/2$ of their claims; each additional dollar of estate is shared equally between all creditors.

Consider now two claimants i and j, where $d_i \leq d_j$. When E is small, they receive equal amounts. This continues until i has received $d_i/2$. Beyond that, i leaves the picture for the time being, and only j may receive any part of each additional dollar; that is, each additional dollar received by both together goes to j. This continues until j has received all but $d_i/2$ of her claim; beyond that, i and j again receive equal shares of each additional dollar. But this is precisely the verbal description of the CG solution given in the previous section. This shows that the solution we have exhibited is indeed consistent, and completes the proof of Theorem A.

Define a *rule* as a function that assigns a solution to each bankruptcy problem. The *CG-consistent* (or simply *consistent*) rule is the one that assigns the CG-consistent solution to each bankruptcy problem. A rule f is called *self-consistent* if

$$f(E;d) = x \quad \text{implies} \quad f(x(S);d|S) = x|S$$

for each set S of creditors, where $x|S$ means "x restricted to S," and $x(S)$ is short for $\sum_{i \in S} x_i$. In words, *any* subset S of the set of all creditors (not only a 2-person subset) uses the rule f to divide among its members the total amount $x(S)$ that it gets when the rule f is applied to the original bankruptcy problem.

Corollary 3.1. *The CG-consistent rule is self-consistent.*

Proof. Let $(E;d)$ be a bankruptcy problem, x its CG-consistent solution, g the function that assigns to each 2-person bankruptcy problem its CG solution, and S a set of creditors. For *any* i, j, the CG-consistency of x yields $(x_i, x_j) = g(x_i + x_j; d_i, d_j)$; in particular, this is so for $i, j \in S$. But that means that $x|S$ is the CG-consistent solution of $(x(S);d|S)$.

Self-consistency and CG-consistency are totally different kinds of concepts. Self-consistency applies to *rules*, CG-consistency to *individual solutions*. If three creditors come to a judge and ask him to divide an estate between them, and he does so in some specific way, then they cannot complain that he is not self-consistent; to do that they would have to know what he would have decided in other situations. But CG-consistency can be checked directly for each proposed solution of each case separately. If one thinks of the principle of CG-consistency as "just," then one can complain about the injustice of one particular decision; the corresponding statement cannot be made for self-consistency.

The CG-consistent rule is by no means the only self-consistent one. Others include division in proportion to the claims, and the constrained equal division solutions in the next section; these rules also play an important role in the Talmudic discussion of bankruptcy.

Various consistency conditions that are similar in spirit to those discussed here play an important role in game theory and bargaining theory. For examples, see Sections 9.6 and 9.7 below.

9.4. Self-Duality and Constrained Equal Division

Define the *dual* f^* of a rule f by

$$f^*(E;d) := d - f(D - E;d);$$

f^* assigns awards in the same way that f assigns losses. A *self-dual* rule is one with $f^* = f$; such a rule treats losses and awards in the same way.

Two prominent features of the consistent rule, both of which follow from the explicit characterization in the proof of Theorem A, are its self-duality, and the qualitative change in the rule that occurs at $E = D/2$. In this section we discuss these features and show that they are strongly rooted in the Talmudic literature. We end the section with an alternative characterization (Theorem B) of the consistent rule in terms of these features, a characterization not directly related to the CG principle.

The basic idea behind duality is that there are certain types of division problem in which it is natural to think in terms of dividing the award (amount received, gain), and other problems in which it is more natural to think in terms of dividing the loss. All other things being equal, it seems appropriate to apply dual solution rules to these problems. In still other problems, it is equally natural to think in terms of losses or of gains; in such cases a self-dual rule is called for.

We start by illustrating the notion of duality with some rules other than the consistent one. A *constrained equal award* (CEA) solution of a bankruptcy problem (E; d) is one of the form $(\alpha \wedge d_1, \ldots, \alpha \wedge d_n)$, where $a \wedge b := \min(a, b)$. In words, this means that all claimants get the same award α, except that those who claim less than α get their claims. Note that

each bankruptcy problem has a unique CEA solution. (4.1)

Indeed, $\sum_{i=1}^{n} \alpha \wedge d_i$ is a continuous strictly increasing function of α on the interval $[0, d_n]$, and maps this interval onto $[0, D]$; hence every point in $[0, D]$ is attained precisely once, proving (4.1).

In our Mishna, the CEA rule prescribes equal awards to all creditors up to $E = 300$; it prescribes (100, 150, 150) for $E = 400$, and

(100, 200, 200) for $E = 500$. The rule, which divides each additional dollar equally between those claimants who still have an outstanding claim, seems natural enough; it has been adopted as law by most major codifiers, including Maimonides[12] (1135–1204).

Maimonides's great intellectual adversary, Rabad,[13] adopted a different rule. This provides equal awards for all creditors when $E \leq d_1$. When E passes d_1, 1 leaves the picture, and each additional dollar is divided equally among the remaining $n-1$ creditors. When E passes d_2, also 2 leaves the picture, and each additional dollar is divided equally among the remaining $n-2$ creditors; and so on. The rule is not defined beyond $E = d_n$. This, too, seems quite natural; when $E \leq d_n$ one might say that all creditors have a claim on the first d_1 dollars, only $2, \ldots, n$ on the next $d_2 - d_1$ dollars, and so on.[14]

We would not discuss these rules in such detail if it were not for the remarkable fact that their duals also make an explicit appearance in the Talmud (Erakhin 27b).[15] At an auction there are n bidders, who bid $b_1 < b_2 < \cdots < b_n$. If n reneges, the object is acquired by $n-1$, and the seller sustains a loss of $b_n - b_{n-1}$; this loss must be paid by n, as the price of being allowed out of his contract. Suppose now that all n bidders renege, and that for one reason or another the object cannot be sold to anyone else; then the loss to the seller is b_n, and this must somehow be shared among the bidders. How?

In this case, Maimonides says that the loss is divided equally among all bidders, subject, of course, to no bidder paying more

[12] *The Laws of Lending and Borrowing*, Chapter 20, Section 4.

[13] Acronym for Rabbi Abraham ben David (1125–1198).

[14] This rule is implicit already in the Babylonian Talmud's discussion of our Mishna, and so goes back at least to the third or fourth century. It first appears explicitly in Alfasi (op. cit.), and is also mentioned by Rashi and many other medieval commentators. Only Rabad, though, seems to have adopted it as law; see his gloss on Alfasi (op. cit.). The rule is also mentioned by the mathematician Abraham Ibn Ezra in connection with certain types of inheritance problem [3, p. 60ff.]; cf. Rabinovitch [17, p. 162] and O'Neill [14]. In modern times it has surfaced again as the solution to the airport landing problem [9]; it is closely connected with the Shapley value [19], a game-theoretic solution concept that is conceptually quite different from the nucleolus.

[15] We are grateful to Y. Aumann for bringing this reference to our attention.

than his bid.[16] Rabad,[17] on the other hand, divides the loss into the n successive increments $b_1, b_2 - b_1, b_3 - b_2, \ldots, b_n - b_{n-1}$. The first increment is paid by all bidders in equal shares, the second in equal shares by all except 1, and so on. These rules are the exact duals of the rules adopted by these same authorities for bankruptcy.

It is apparent that these authorities had precise ideas on how to deal with division problems, and that they applied them to the award or the loss according to whether the funds were to be received (as in bankruptcy) or paid (as in the auction). But one can also approach the problem not so much from the technical viewpoint of the direction in which money flows at the specific time of the court decision, but from the more substantive viewpoint of whether the protagonists themselves consider the transaction an award or a loss. In bankruptcy, for example, the creditors will in the end receive checks as a result of the court proceedings. Nevertheless, they are worse off than before making the loan, and they may well conceive of the transaction as a loss rather than an award.

This suggests a rule in which (i) awards and losses are treated dually, and (ii) it makes no difference whether we think of the outcome as an award or a loss.[18] Together, (i) and (ii) call for a self-dual rule.

Self-duality was just one of the two prominent features of the consistent rule mentioned at the beginning of this section. The other was the qualitative change in the rule that occurs at the "halfway point," $E = D/2$. This, too, is strongly rooted in the Talmud; there are dozens — perhaps hundreds — of discussions hinging on the principle that "more than half[19] is like the whole" (Hulin 27a).

[16] *The Laws of Appraisal,* Chapter 8, Section 4. Unlike for bankruptcy, Maimonides here gives no clear general rule; for the specific numerical example treated in the Talmud (and by Maimonides), his division is equal. But it is difficult to imagine that Maimonides would ever require a reneging bidder to pay more than his bid.

[17] See his gloss on Maimonides, op. cit.

[18] Specifically, whether we think of the outcome to Creditor i as an award of x_i or a loss of $d_i - x_i$.

[19] "Rov."

For example, kosher slaughter of an animal calls for cutting through the windpipe and the foodpipe; but as long as more than half of each pipe is cut, the meat is still kosher (op. cit.).

Another example is based on the Talmudic principle that in general, a lender automatically has a lien on the borrower's real property. But when his entire property is worth less than half the loan, the borrower may in certain cases dispose of it "free and clear" (Erakhin 23b). Rashi explains that when the property is worth more than half the loan, the lien is of considerable importance, and the lender relies on it as a guarantee. But when it is worth less than half the loan, the lien will not help very much anyway; the loan was presumably made "on trust," and we are not justified in repossessing the property from the bona fide recipient.

Again, the principle involved here is "more than half is like the whole;" property amounting to more than half the loan is conceptually close to covering it all, and cannot be ignored. Less than half is like nothing; property covering less than half the loan is inconsiderable, need not be taken into account. This is not merely a legal convention, but is explicitly based on a psychological presumption. In the "less-than-half" case, the lender is presumed not to rely on the lien, to lend "on trust." Psychological presumptions of this kind often play an important role in Talmudic law.[20]

In the bankruptcy problem, too, the half-way point is a psychological watershed. If you get more than half your claim, your mind focusses on the full debt, and your concern is with the size of your loss. If you get less than half, your mind writes off the debt entirely, and is "happy" with whatever it can get; your concern is with your award. Moreover, it is socially unjust for different creditors to be on opposite sides of this watershed; for one creditor to get most of his claim, while another one loses most of his. Subject to this constraint, therefore, the losses are divided equally when $E \geq D/2$, the awards when $E \leq D/2$. In brief, we have

[20]E.g., the finder's right to found property depends on whether the loser can be presumed to have "despaired" of regaining it (Baba Metzia 21a ff.).

Theorem B. *The consistent rule is the unique self-dual rule that, when $E \leq D/2$, assigns to $(E; d)$ the constrained equal award solution of $(E; d/2)$.*

Mathematically, this theorem is simply a concise expression of the explicit construction in the proof of Theorem A. Conceptually, though, it provides an additional, independent characterization of the consistent rule.

We end this section by mentioning two properties of the consistent rule that will be useful in the sequel. Call a rule f *monotonic* if $f_i(E; d)$ is a non-decreasing function of E when i and d are held fixed; that is, no claimant loses from an increase in the estate. Call a solution x to a bankruptcy problem $(E; d)$ *order-preserving* if

$$0 \leq x_1 \leq x_2 \leq \cdots \leq x_n \quad \text{and} \quad 0 \leq d_1 - x_1 \leq \cdots \leq d_n - x_n.$$

That is, a person with a higher claim than another gets an award that is no smaller and suffers a loss that is no smaller;[21] ordinally, one might say, both the award and the loss are scaled to the claim. As we have said, the consistent rule is monotonic and yields order-preserving solutions. So are the other rules we have considered here: division in proportion to the claims, the CEA rule, its dual, and the solution adopted by Rabad (insofar as it is defined).

9.5. Coalition Formation and the Jerusalem Talmud

In discussing our Mishna, the Jerusalem Talmud[22] says as follows: "Samuel says, the Mishna takes it that the creditors empower each other; specifically, that the third empowers the second to deal with

[21] Also, awards are non-negative and do not exceed the claims; but this statement follows from the other if we include outsiders, whose claim is 0.

[22] The Jerusalem Talmud is based on the same source (the Mishna) as the Babylonian Talmud, and is contemporaneous with it. Although considered less authoritative, it is valued as an independent parallel source, which often sheds light on obscure passages in the Babylonian Talmud. We are very grateful to Yehonatan Aumann for calling to our attention the remarkable passage that forms the basis for this section.

the first. She may say to her, 'Your claim is 100, right? Take 50 and go.' "

Samuel is obviously referring to the cases $E = 200$ and $E = 300$. The second and third creditors (whose claims are 200 and 300, respectively) form a coalition "against" the first (whose claim is 100). This leaves two effective protagonists, with claims of 500 and 100, respectively; applying the contested garment rule yields the first creditor 50, with the remainder going to the coalition. If the coalition again applies the contested garment rule to divide its award among its members, the numbers in the Mishna result.

If applied to the case $E = 100$, this procedure would lead to the payoff vector (50, 25, 25), which is not order-preserving: 1's award is larger than that of 2 or of 3. It begins yielding order-preserving results at $E = 150$; as E rises, it continues to do so until $E = 450$. Beyond that, order preservation again fails, this time on the loss side. At $E = 500$, for example, we obtain (50, 175, 275), so that l's loss is larger than that of 2 or of 3.

We may proceed in the same way in any problem with three creditors. First 2 and 3 pool their claims and act as a single agent vis-a-vis 1. The CG solution of the resulting problem yields awards to 1, and to the coalition of 2 and 3; to divide its award among its members, the coalition again applies the CG principle. The result is order preserving if and only if $3d_1/2 \leq E \leq D - (3d_1/2)$. If one divides the awards equally when $E \leq 3d_1/2$, and the losses equally when $E \geq D - (3d_1/2)$, one obtains precisely the consistent solution over the entire range $0 \leq E \leq D$.

By using induction, one may generalize this in a natural way to arbitrary n. Suppose we already know the solution for $(n-1)$-person problems. Depending on the values of E and d, we treat a given n-person problem in one of the following three ways:

(i) Divide E between $\{1\}$ and $\{2, \ldots, n\}$ in accordance with the CG solution of the 2-person problem $(E; d_1, d_2 + \cdots + d_n)$, and then use the $(n-1)$-person rule, which we know by induction, to divide the amount assigned to the coalition $\{2, \ldots, n\}$ between its members.

(ii) Assign equal awards to all creditors.
(iii) Assign equal losses to all creditors.

Specifically, (i) is applied whenever it yields an order-preserving result, which is precisely when $nd_1/2 \leq E \leq D - (nd_1/2)$. We apply (ii) when $E \leq nd_1/2$, and (iii) when $E \geq D - (nd_1/2)$. This is called the *coalitional procedure.*

For example, let $n = 5$, $d_i = 100\ i$, $E = 510$. At the first step of the induction, the coalition $\{2, 3, 4, 5\}$ forms; its joint claim is 1400, while l's claim is 100. Applying the CG rule yields 50 to 1, and 460 to the coalition. To divide the 460 between 2, 3, 4, and 5, one splits $\{2, 3, 4, 5\}$ into 2 and $\{3, 4, 5\}$, and again applies the CG rule. This yields 100 to 2, and 360 to $\{3, 4, 5\}$. If one were again to split $\{3, 4, 5\}$ into 3 and $\{4, 5\}$, then 3 would be awarded 150, leaving 4 and 5 only 210 to divide between them. At least one of them would therefore receive ≤ 105, so that the result would not be order-preserving. At this point, therefore, the 360 are split equally between 3, 4, and 5. The final result is (50, 100, 120, 120, 120), which *is* order-preserving.

Note that this is the consistent solution of the above problem. More generally, it may be verified that

Theorem C. *The coalitional procedure yields the consistent solution for all bankruptcy problems.*

Theorems A and C both use the CG principle to characterize the same rule; but the two characterizations are conceptually totally different. Theorem A applies the CG principle to pairs of *individuals* only, and it is applied to all $\binom{n}{2}$ such pairs. The $\binom{n}{2}$ resulting conditions are desiderata of a solution, but they do not tell us directly how we should arrive at one; it is not *a priori* clear that they have any simultaneous solution at all, or that they do not have more than one. Theorem C applies the CG principle to pairs of *coalitions*; and it describes an orderly step-by-step process, which by its very definition must lead to a unique result. But it uses only certain carefully selected pairs of coalitions, not all such pairs.[23]

9.6. The Nucleolus and the Kernel

It will be recalled that a (*coalitional*) *game* is a function v that associates a real number $v(S)$ with each subset S of a finite set N. The members of N are called *players*, the sets S *coalitions*. Intuitively, $v(S)$ represents the total amount of payoff that the coalition S can get by itself, without the help of other players; it is called the *worth* of S. By convention, $v(\varnothing) = 0$. A *payoff vector* is a vector x with components indexed by the players; x_i represents the payoff to i.

Solution concepts associate payoff vectors with games; each such concept represents a specific notion of stability, expected outcome, or the like. In many cases a solution concept associates several payoff vectors with a game, or none at all. Only two of the better known solution concepts associate a unique payoff vector with each game; they are the value [19] and the nucleolus [18].

As it stands, the bankruptcy problem considered here is not a game; coalitions do not appear explicitly in its formulation. A natural way to associate a game with a bankruptcy problem $(E;\, d)$ is to take the worth of a coalition S to be what it can get without going to court; i.e., by accepting either nothing, or what is left of the estate E after each member i of the complementary coalition $N \backslash S$ is paid his complete claim d_i. Thus we define the (*bankruptcy*) *game* $v_{E;d}$ *corresponding* to the bankruptcy problem $(E;\, d)$ by

$$v_{E;d}(S) := (E - d(N \backslash S))_+. \tag{6.1}$$

[23] As described — with the coalitions $\{1\}$ and $\{2, \ldots, n\}$ — the coalitional procedure yields a monotonic rule and order-preserving solutions. Moreover, it appears that they are the *only* such coalitions, though we have no satisfactory formulation and proof of such a result. Also, it appears that if the creditors may form coalitions as they wish, then for $E \leq D/2$, the incentives lead to the coalitions suggested by the coalitional procedure. These matters call for further study.

Theorem D. *The consistent solution of a bankruptcy problem is the nucleolus of the corresponding game.*

The proof of Theorem D makes use of several concepts and results of cooperative game theory. Let v be a game, S a coalition, x a payoff vector. The *reduced game* $v^{S,x}$ is defined [1, 20, 10, 15] on the player space S as follows:

$$\begin{aligned} v^{S,x}(T) &= x(T) && \text{if } T = S \text{ or } T = \varnothing, \\ &= \max\{v(Q \cup T) - x(Q) : Q \subset N \backslash S\} && \text{if } \varnothing \subsetneqq T \subsetneqq S. \end{aligned}$$

In the reduced game, the players of S consider how to divide the total amount assigned to them by x under the assumption that players i outside S get exactly x_i. Together, all the players of S get $x(S)$; as always, the empty set gets nothing. If a non-empty proper subcoalition T of S chooses a set Q of "partners" outside S, it will have total worth $v(Q \cup T)$; but to keep the partners satisfied, it will have to pay them the total $x(Q)$ assigned to them by x. Thus T will choose its partners Q to maximize the amount $v(Q \cup T) - x(Q)$ left for it after paying off the partners.

Lemma 6.2. *Let x be a solution of the bankruptcy problem $(E; d)$, such that $0 \leq x_i \leq d_i$ for all i. Then for any coalition S,*

$$v_{E;d}^{S,x} = v_{x(S);d|S}. \tag{6.2}$$

(In words, the reduced bankruptcy game is the game corresponding to the "reduced bankruptcy problem.")

Proof. Set $v := v_{E;d}$ and $v^S := v_{E;d}^{S,x}$. First let $\varnothing \subsetneqq T \subsetneqq S$, and let the maximum in the definition of $v^S(T)$ be attained at Q. Since $x_i \geq 0$ and $a_+ - b_+ \leq (a-b)_+$ for all a and b, we have

$$\begin{aligned} v^S(T) &= v(T \cup Q) - x(Q) = (E - d(N \backslash (Q \cup T)))_+ - (x(Q))_+ \\ &\leq (x(N) - d(N \backslash (Q \cup T)) - x(Q))_+ \\ &= [x(S) - d(S \backslash T) - (d - x)(N \backslash (S \cup Q))]_+ \\ &\leq (x(S) - d(S \backslash T))_+, \end{aligned} \tag{6.3}$$

where the last inequality follows from $x_i \leq d_i$. On the other hand, setting $Q = N \backslash S$ yields

$$\begin{aligned} v^S(T) &\geq v(T \cup (N\backslash S)) - x(N\backslash S) \\ &= (E - d(N\backslash(T \cup (N\backslash S))))_+ - (x(N) - x(S)) \\ &\geq (E - d(S\backslash T)) - (E - x(S)) = x(S) - d(S\backslash T); \end{aligned} \tag{6.4}$$

and setting $Q = \varnothing$ yields

$$v^S(T) \geq v(T \cup \varnothing) - x(\varnothing) = v(T) = (E - d(N\backslash T))_+ \geq 0. \tag{6.5}$$

Formulas (6.4) and (6.5) together yield

$$v^S(T) \geq (x(S) - d(S\backslash T))_+;$$

together with (6.3), this yields

$$v^S(T) \geq (x(S) - d(S\backslash T))_+ = v_{x(S);d|S}(T). \tag{6.6}$$

When $T = \varnothing$ or $T = S$, formula (6.6) is immediate, so the proof of the lemma is complete.

We also make use of the solution concepts called kernel [1] and pre-kernel [10]. Let v be a game. For each payoff vector x and players i, j, define

$$s_{ij}(x) = \max\{v(S) - x(S)\colon S \text{ contains } i \text{ but not } j\}.$$

The *pre-kernel* of v is the set of all payoff vectors x with $x(N) = v(N)$ and $s_{ij}(x) = s_{ji}(x)$ for all i and j. The *kernel* of v is the set of all payoff vectors x with $x(N) = v(N), x_i \geq v(i)$ for[24] all i, and for all

[24]We do not distinguish between i and $\{i\}$.

i and j,

$$s_{ij}(x) > s_{ji}(x) \quad \text{implies} \quad x_j = v(j).$$

One more definition is required. The *standard solution* of a 2-person game v with player set $\{1, 2\}$ is given by

$$x_i = \frac{v(12) - v(1) - v(2)}{2} + v(i). \tag{6.7}$$

Note that this is equivalent to $x_1 + x_2 = v(12)$, $x_1 - x_2 = v(1) - v(2)$. In words, the standard solution gives each player i the amount $v(i)$ that he can assure himself, and divides the remainder equally between the two players. The nucleolus and kernel,[25] the pre-kernel and the Shapley value of a 2-person game all coincide with its standard solution; so do most of the better-known bargaining solutions [13, 4, 12]. Indeed, the standard solution constitutes the only symmetric and efficient point-valued solution concept for 2-person games that is covariant under strategic equivalence.[26]

Lemma 6.8. *Let x be in the pre-kernel of a game v, and let S be a coalition with exactly two players. Then $x|S$ is the standard solution of $v^{S,x}$.*

Proof. Let $S = \{i, j\}$. Then

$$\begin{aligned} s_{ij}(x) &= \max_{Q \subset N \setminus S} (v(Q \cup i) - x(Q \cup i)) \\ &= \max_{Q \subset N \setminus S} (v(Q \cup i) - x(Q)) - x_i = v^{S,x}(i) - x_i; \end{aligned}$$

similarly $s_{ji}(x) = v^{S,x}(j) - x_j$. Since by the definition of pre-kernel, $s_{ij}(x) = s_{ji}(x)$, it follows that $v^{S,x}(i) - x_i = v^{S,x}(j) - x_j$. Hence

$$x_i - x_j = v^{S,x}(i) - v^{S,x}(j)$$

[25] When they are non-empty, which is the case whenever there is at least one payoff vector x that is both individually rational ($x_i \geq v(i)$ for all i) and efficient ($x(N) = v(N)$).

[26] A similar remark is made in [15]. Rules such as proportional division or the CEA rule are not covariant in terms of the game v.

and

$$x_i + x_j = x(i, j) = v^{S,x}(i, j),$$

which proves the lemma.[27]

Remark. The converse of this lemma is also true; i.e., if $x(N) = v(N)$ and $x|S$ is the standard solution of $v^{S,x}$ for all 2-person coalitions S, then x is in the pre-kernel of v. From this it follows that if $|N| \geq 3$, then x is in the pre-kernel of v if and only if $x(N) = v(N)$ and $x|S$ is in the pre-kernel of $v^{S,x}$ for all coalitions $S \subsetneqq N$ [14].

Lemma 6.9. *The contested garment solution of a 2-person bankruptcy problem is the standard solution of the corresponding game.*

Proof. Follows from (2.1), (6.1), and (6.7).

Proposition 6.10. *The kernel of a bankruptcy game $v_{E;d}$ consists of a single point, namely the consistent solution of the problem (E;d).*

Proof. Set $v = v_{E;d}$, and let x be in its kernel. By its definition (6.1), v is superadditive ($S \cap T = \varnothing \Rightarrow v(S) + v(T) \leq v(S \cup T)$) and hence 0-monotonic ($S \subset T \Rightarrow v(S) + \sum_{i \in T \setminus S} v(i) \leq v(T)$). In 0-monotonic games, the kernel coincides with the pre-kemel [11]. Hence x is in the pre-kemel of v.

Now let S be an arbitrary 2-person coalition. By Lemma 6.8, $x|S$ is the standard solution of $v^{S,x}$, and hence by Lemma 6.2, of $v_{x(S);d|S}$. Hence by Lemma 6.9, $x|S$ is the CG-solution of $(x(S); d|S)$; but that means that x is the consistent solution of $(E;\ d)$.

Theorem D follows from Proposition 6.10, since the nucleolus is always in the kernel [18].[28]

We stated Theorem D in terms of the nucleolus because it is better known than the kernel and conceptually simpler (it is point

---

[27] This lemma is a special case of Lemma 7.1 in [1].

[28] Since this is the only property of the nucleolus that we require, we will not cite its definition here. The interested reader may consult the original article [18], or any one of several equivalent characterizations [5, 6, 20].

valued). In fact, though, the idea of CG-consistency is more closely related to the kernel than it is to the nucleolus.[29] We have already noted that the contested garment solution is in fact simply the standard solution of 2-person games. Thus an appropriate generalization of the notion of a consistent solution to an arbitrary game v is a payoff vector x such that whenever S is a 2-person coalition, $x|S$ is the standard solution of the reduced game $v^{S,x}$. Lemma 6.8, together with the theorem of Peleg cited in the succeeding remark, show that with this definition, the set of all consistent solutions of an arbitrary game is precisely its pre-kernel; and as we have noted, this coincides with the kernel for 0-monotonic games.

9.7. Historical Notes

In the Talmudic bankruptcy literature, the rule that is perhaps closest to ours was proposed by Piniles [16, p. 64]; it coincides with ours for $E \leq D/2$, but beyond that they differ. Evidently, Piniles was unaware of the connection with the contested garment, since even for the CG Mishna itself (where $E = 1 > \frac{3}{4} = D/2$), his rule gives $(\frac{5}{8}, \frac{3}{8})$ rather than $(\frac{3}{4}, \frac{1}{4})$. We owe the reference to Erakhin 23b (see Section 9.4) to Piniles.

In bargaining theory, the idea of consistency first appeared in [2, p. 328], where Harsanyi characterized the product maximization solution to an n-person bargaining problem as the unique solution x at which each two players i, j use the Nash solution [13] to divide between them what remains if every other player k gets x_k.

In the context of apportionment, self-consistency is discussed in Balinski and Young's "Fair Representation" (Yale University Press, 1982, 43–45 and 141–149). Alexander Hamilton's apportionment method, vetoed by George Washington but nevertheless used in the U.S. from 1852 until 1901 (and in many countries until today), is not self-consistent. Just before Oklahoma became a state in 1907, the House of Representatives had 386 seats. Oklahoma was allocated 5,

[29] The nucleolus has more to do with self-consistent rules (Section 9.3) than with CG-consistent solutions. See [20].

bringing the total to 391. Nothing else changed, so presumably the apportionment among the old states should have remained the same. Yet under Hamilton's method, Oklahoma's joining meant New York losing a seat to Maine!

Others who have recently used ideas related to consistency include H. Moulin and W. Thomson. A particularly striking result is by T. Lensberg [7], who showed that Nash's bargaining solution [13] can be characterized by a set of axioms in which the Independence of Irrelevant Alternatives is replaced by a self-consistency axiom.

Acknowledgments

Particular thanks are due to Y. Aumann, Y. Kannai, D. Kleitman, and B. Peleg, both for the specific contributions acknowledged in the text and footnotes, and for lengthy discussions on various other aspects of this research. We have also benefitted from discussions with many other individuals. What started us on this research was reading B. O'Neill's beautiful paper [14], and then being led to the Mishna in Kethubot 93a as the result of a correspondence on the subject of O'Neill's paper with the late S. Aumann.

This work was supported by National Science Foundation Grant SES 83-20453 at the Institute for Mathematical Studies in the Social Sciences, Stanford University, and by the Institute for Mathematics and its Applications at the University of Minnesota.

References

Note. Items from the ancient and medieval Talmudic literature are not listed here; citations of this literature in the text and footnotes use the commonly accepted style.

1. M. Davis and M. Maschler, The kernel of a cooperative game, *Naval Res. Logist. Quart.* **12** (1965), 223–259.
2. J. Harsanyi, A bargaining model for the cooperative n-person game, *in* "Contributions to the Theory of Games IV," pp. 325–355, Annals of Mathematics Studies Vol. 40, Princeton Univ. Press, Princeton, N.J., 1959.
3. A. Ibn Ezra, "Sefer Hamispar" ("The Book of the Number"), Verona, 1146; German translation by M. Silberberg, Kaufimann, Frankfurt a. M., 1895. [Hebrew]

4. E. Kalai and M. Smorodinski, Other solutions to Nash's bargaining problem, *Econometrica* **43** (1975), 513–518.
5. E. Kohlberg, On the nucleolus of a characteristic function game, *SIAM J. Appl. Math.* **20** (1971), 62–66.
6. E. Kohlberg, The nucleolus as a solution to a minimization problem, *SIAM J, Appl. Math.* **23** (1972), 34–39.
7. T. Lensberg, Stability and the Nash Solution, *J. Econ. Theory* **45** (1988), 330–341.
8. I. Lewy, Interpretation des IV. Abschnittes des palast. Talmud-Traktats Nesikin (Commentary on the Fourth Chapter of the Tractate Nezikin of the Jerusalem Talmud), *Jahr. Judischen Theologischen Sem. Breslau* (1908), 101–131. [Hebrew]
9. S. C. Littlechild and G. Owen, A simple expression for the Shapley value in a special case, *Manage. Sci.* **20** (1973), 370–372.
10. M. Maschler, B. Peleg, and L. S. Shapley, The kernel and bargaining set for convex games, *Int. J. Game Theory* **1** (1972), 73–93.
11. M. Maschler, B. Peleg, and L. S. Shapley, Geometric properties of the kernel, nucleolus, and related solution concepts, *Math. Oper. Res.* **4** (1979), 303–338.
12. M. Maschler and M. Perles, The super-additive solution for the Nash bargaining game, *Int. J. Game Theory* **10** (1981), 163–193.
13. J. F. Nash, The bargaining problem, *Econometrica* **18** (1950), 155–162.
14. B. O'Neill, A problem of rights arbitration from the Talmud, *Math. Soc. Sci.* **2** (1982), 345–371.
15. B. Peleg, On the reduced game property and its converse, *Int. J. Game Theory* **15** (1986), 187–200.
16. H. M. Piniles, "Darkah shel Torah" ("The Way of the Law"), Forster, Vienna, 1861. [Hebrew]
17. N. L. Rabinovitch, "Probability and Statistical Inference in Ancient and Medieval Jewish Literature," Univ. Toronto Press, Toronto, Buffalo, 1973.
18. D. Schmeidler, The nucleolus of a characteristic function game, *SIAM J. Appl. Math.* **17** (1969), 1163–1170.
19. L. S. Shapley, A value for n-person games, *in* "Contributions to the Theory of Games II," pp. 307–312, Annals of Mathematics Studies Vol. 28, Princeton Univ. Press, Princeton, 1953.
20. A. I. Sobolev, Charakterizatziya Printzipov Optimalinosti v Kooperativnych Igrach Posredstvom Funktzionalnych Uravneniy (Characterization of the Principle of Optimality for Cooperative Games through Functional Equations), *in* "Matematicheskie Metody v Socialnich Naukach" (N. N. Vorobiev, Ed.), pp. 94–151, Vipusk 6, Vilnius, 1975. [Russian]

Chapter 10

Game Theoretic Aspects of Gradual Disarmament*

10.1. Introduction

The task that was assigned to this group concerns the application of utility theory to disarmament, and the study here described is addressed to that task. However, the implications of this study range far beyond the area of disarmament and touch on all areas of negotiations and international relations in general.

Before starting, let us express at once what we think can be a role of mathematical studies and mathematical model building in the realm of international politics. Undoubtedly, many questions of international politics are too ramified to be directly amenable to exact mathematical treatment. No mathematical treatment can possibly take account of all the implications involved in a political decision. One therefore builds a simplified mathematical model, which is an abstraction that treats in highly simplified form just one or a few of the aspects of a complex situation. The analysis of such a highly simplified abstraction can very seldom lead to any specific recommendations in a specific situation. But it can lead to insights of a general nature. These insights can then be used by policy makers in making specific decisions or in formulating general policies.

---

*Co-authored with M.B. Maschler. Reprinted from *Repeated Games with Incomplete Information* (1995), Chapter 1, 1–41, with permisson of The MIT Press.

Thus, for example, an analysis of a simplified model may direct attention to a new procedure not previously discovered because of the complexity of the actual problem. It is this kind of use of mathematical models that concerns us here.

Needless to mention, the relation between the mathematical model and the real situation works in both directions: It is the real situation which provides hints as to how the model should be constructed; it is the mathematical analysis of the model which directs attention to new procedures which can be used by the policy makers; and it is the attempt to consider the implications of these procedures in reality which leads to a new and more sophisticated model for a further study.

In any bargaining situation, it is of the utmost importance for each side to know how the other side evaluates the situation. Consider, for example, the following famous problem: Two brothers, Tom and Dick, must share a pie in a "fair" manner between them. A reasonable procedure would be that one of them should divide the pie into two parts and let the other one choose one of the parts. Clearly, if the divider cuts the pie into two parts that seem to him to be of equal size, then each brother is satisfied at the end that he has received at least half the pie, which is fair enough.

Consider now a pie with a cherry. Suppose Tom is indifferent to cherries; namely, does not care whether there is a cherry in his part or not; and suppose that Tom knows that Dick likes cake but also knows that Dick loves cherries. Using the above procedure he would then divide the cake into unequal portions, leaving the cherry in the smaller part, knowing that his brother would pick up this part — leaving himself with a greater portion of the cake. How small the portion with the cherry should be depends, of course, on Tom's knowledge of the amount of cake that Dick would be willing to sacrifice for the cherry.

Clearly, it is advantageous for Dick to hide his preference for cherries, or at least to give the impression that the preference is small.

Consider now a case in which Tom is indifferent to cherries (and Dick knows this) and Dick loves cherries but Tom does not know this, although he may suspect it. (Dick knows that Tom does not know his preferences.) Suppose, however, that both brothers also know that they will have to divide a cake every day during many days. It stands to reason that Tom should test his brother's preferences by cutting the cake into different sizes during some of the days and check which part has been chosen. If Dick would always pick up that part which he really prefers, he will reveal in a few days his trading price; and henceforth Tom would be able to take full advantage of the knowledge thus acquired.

Should Dick act this way? Should he act as if he does not love cherries? Should he choose the smallest portion in a random way, according to a fixed probability? Or should he act in still another way?

These questions are not frivolous: the kind of considerations they typify are basic for disarmament negotiations. Suppose, for example, that a gradual disarmament procedure is proposed, under which both the United States and the Soviet Union propose gradually to destroy a given collection of weapons. One way of doing so is to let each country divide its own remaining collection each year into, say, ten parts and let the other country choose which part should be destroyed. Clearly, a "naïve" way of choosing may reveal secret military preferences and advantage can be taken of this when the remainder is to be divided, not to mention other political and military advantages that may accrue. How, then, should one choose in an optimal way?

Our object in this chapter is to study some very simple examples — examples that can be solved explicitly — and point out several interesting features of the solutions. We shall also describe a general result concerning the solution of a class of such games.[1]

[1]This chapter should be regarded as a heuristic introduction to the subject.

10.2. A Case When a Player Cannot Benefit in the Long Run from His Information

Consider the following two games:

Player 1 \ Player 2	L	R
L	1	0
R	0	0

G_A

Player 1 \ Player 2	L	R
L	0	0
R	0	1

G_B

In each game, Player 1 has two strategies, L and R, and Player 2 has two strategies, L and R. The numbers in the boxes represent the amount that Player 2 must pay Player 1 after the strategies have been chosen (each player makes his choice without knowing the other player's choice). Clearly, the values w_A and w_B of G_A and G_B are zero; i.e., Player 1 can guarantee himself 0 in each game and Player 2 can guarantee not to pay more than 0.

Suppose now that one of these two games is played, but that Player 2 is not informed which one, although he does know that he is playing one of these games. Furthermore, suppose Player 2 believes that the probability that he is playing G_A is 1/2. Player 1, however, knows exactly which game he is playing. Can he use this knowledge to his advantage?

In his theory of games with incomplete information, J. Harsanyi [1967–8] recommends and justifies the following procedure for both players:

Consider the 2-person constant-sum game Γ_1 described by[2] Figure 10.1. (The left branch of this tree represents G_A, and the right branch represents G_B.) It is quite easy to see that the only optimal strategy for Player 1 in Γ_1 is: *Choose "left" if you are at* A, and *choose "right" if you are at* B. For Player 2, any pure or mixed strategy is optimal. The value of the game Γ_1 is 1/2.

So much for Γ_1; but how should one play in the original game, which is G_A or G_B? Harsanyi recommends that Player 2 should play

[2] We use the convention that information sets are indicated by dotted lines.

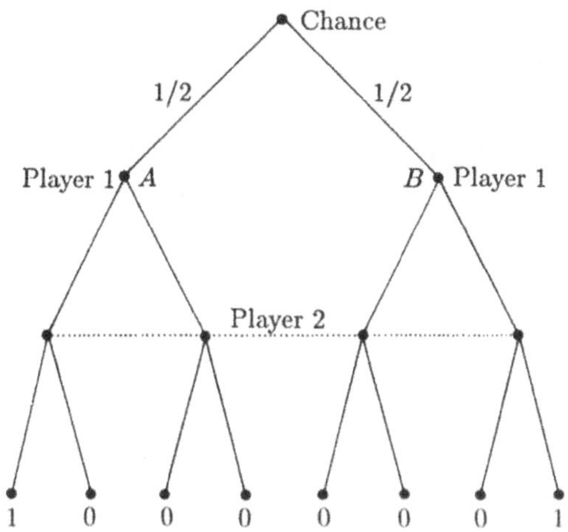

Figure 10.1: The game Γ_1.

in the original game the same optimal strategy as in Γ_1 — in this particular case any pure or mixed strategy he wishes. For Player 1 in the original game he recommends that if G_A is being played (which Player 1 knows), Player 1 should play his optimal strategy for Γ_1 *assuming that he has already reached the vertex* A in the figure; namely — "left". Similarly, if G_B is being played — Player 1 should choose "right".

In order to shorten our description, let us refer to Player 1 when G_A is being played as *type* A of Player 1, or simply Player 1_A. Similarly, type B of Player 1, or simply 1_B, is Player 1 playing the original game when G_B is being played. The term "Player 1" will be reserved for Player 1 playing Γ_1 only. No distinction will be made for Player 2 since he does not know whether G_A or G_B is the true game, and so must play the same way in both cases. With these notations we can summarize the outcome as follows:

i. Player 1 can assure himself 1/2.
ii. The payoffs to 1_A and 1_B depend on the strategy of Player 2. If Player 2 chooses left with probability λ, where $0 \leq \lambda \leq 1$, then 1_A's expected payoff is $v_A^1 = \lambda$ and 1_B's expected payoff is $v_B^1 = 1 - \lambda$.

Note that $v_A^1 \geq w_A, v_B^1 \geq w_B$, and at least one of these inequalities is sharp. Thus, either 1_A or 1_B, and usually both types, are able to take advantage of Player 2's ignorance.

The situation becomes much more complicated if, after the choice of chance, G_A or G_B is played infinitely many times. In other words — if the same game, say G_A, is repeated. Player 1_A knows that G_A is being played but Player 2 does not. His *a priori* subjective belief is 1/2 for G_A and 1/2 for G_B. Let us call a single play of, say, G_A a *stage* in this grand game. At the end of each stage both players are informed of the choices made by their opponents at that stage. Also, at the end of each stage — the appropriate payoff is put aside to the player's credit (debit — if the payoff is negative), but the players are not informed of the amounts.[3] Of course, 1_A can calculate his own payoff since he knows the strategies and the game. Both players want to maximize their *average* expected payoff; i.e., their expected payoff per game.

Harsanyi's model for this game — call it Γ_∞ — is an infinite tree whose first two stages are given in Figure 10.2. Here the appropriate payoffs are written at the end of each stage. Player 1 wants to maximize the average payoff and Player 2 wants to minimize this average. Player 2 has four information sets at the second stage, each containing four vertices.

How should Player 1 play Γ_∞? Clearly it is not advisable to play each stage in the same fashion as in Γ_1 (left on the A-part, right on the B-part). If this were the optimal way of playing, then Player 2 would know it. He would then infer from the choices at the first stage what chance's choice was. From the second stage on he would then be able to force w_A or w_B, i.e., zero payoff for each stage, which

[3]Playing a game without knowing the exact payoff is the usual case in international relations. If, say, there is an agreement under which the Soviet Union destroys 100 I.C.B.M.'s, then if these are all the I.C.B.M.'s they have, the payoff to the United States caused by the destruction is high (from the military point of view); but if the Russians have 1000 I.C.B.M.'s, then the payoff is negligible. Intelligence information may yield some information, but it may be wrong. At best, one can assign subjective probabilities to the various possibilities.

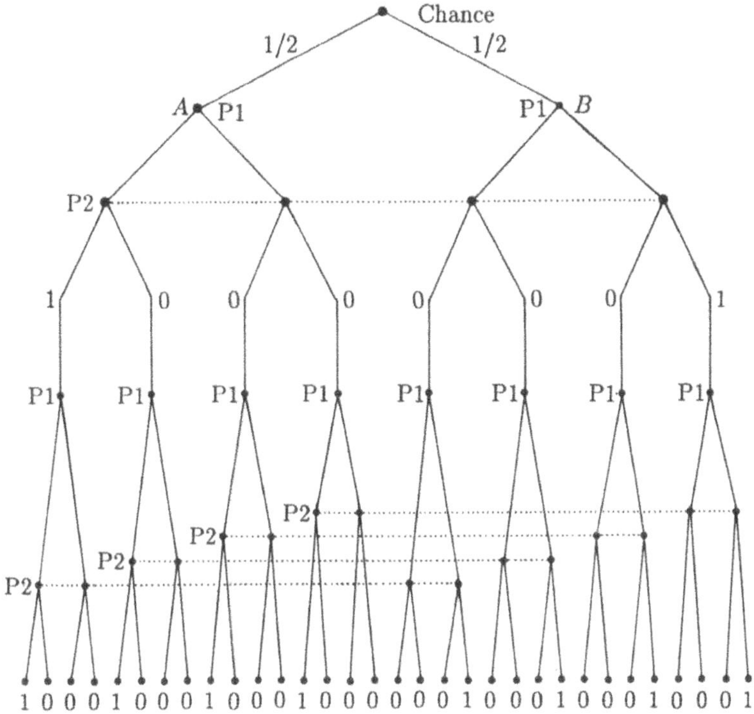

Figure 10.2: Two stages of Γ_∞.

would make the average expected payoff for Player 1 tend to zero as the number of stages tends to infinity.

But Player 1 can assure himself more than zero in Γ_∞. For example, one way for him to do so is to act as if he himself does not know the choice made by chance. In other words, one option for Player 1 is to regard himself as playing an infinite game Δ_∞ whose first two stages are drawn in Figure 10.3.

It turns out that playing Δ_∞ is equivalent to playing the 1-stage game Δ_1 of Figure 10.4 infinitely many times.[4] Game Δ_1 in normal

[4]Clearly, an optimal strategy for each player in Δ_∞ is to play a Δ_1-optimal strategy at each stage.

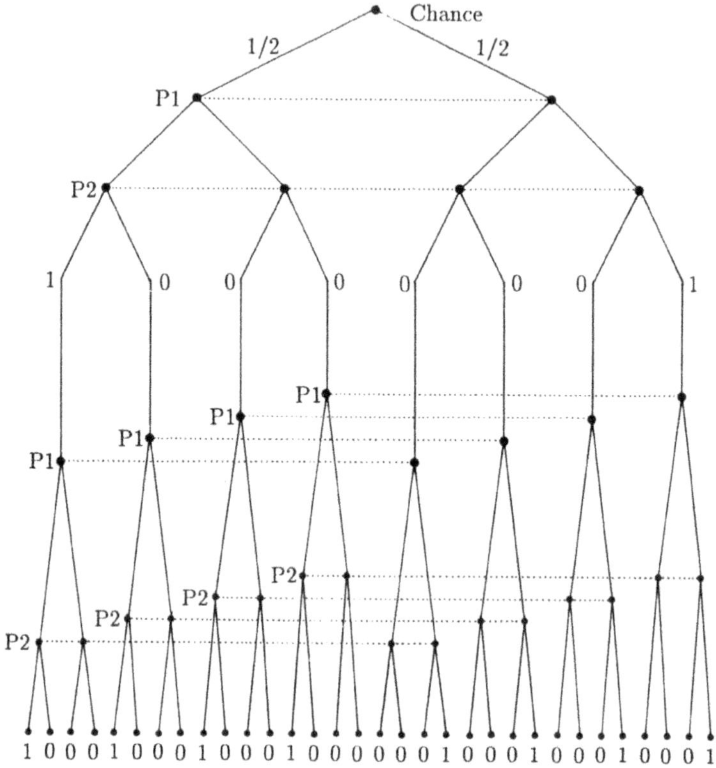

Figure 10.3: Two stages of Δ_∞.

form is represented by the matrix

Player 2

Player 1		L	R
	L	1/2	0
	R	0	1/2

and optimal strategies for it are for each player to choose left with probability 1/2 and right with the same probability. This yields the value $u_1 = 1/4$.

As indicated in the last footnote, a value for the game Δ_1 repeated several times or infinitely often exists and it, too, is $u = 1/4$.

Moreover, optimal strategies for each player are to play each stage as a 1-stage game; namely, to choose left with probability 1/2 and right with the same probability.

Suppose now that Player 1 plays this very same strategy in Γ_∞ instead of in Δ_∞; then he would still assure himself a payoff of 1/4. Indeed, all he does in playing Γ_∞ is to ignore his information on chance's choice, so that his way of playing does not reveal any information on the choice of chance.

We have seen that ignoring his information while playing Γ_∞ assures Player 1 of 1/4. We have also seen that using that information for a 1-stage game yields the value 1/2; but repeating the same optimal strategy at each stage of Γ_∞ is bad. Could Player 1 play better in Γ_∞ by using his information on the choice of chance to guarantee himself an average expected payoff higher than 1/4 (of course less than 1/2)?

The surprising answer is that he cannot — *the value* v_∞ *of* Γ_∞ *is exactly* 1/4. The proof of this statement is relatively complicated; it will be given in Section 10.5. Thus, in this particular case Player 1 cannot benefit in the infinite game from knowing the choice of chance. His optimal strategy is to ignore his knowledge, i.e., play as if he were playing optimally in Δ_∞. We shall see later that any strategy of Player 1 which takes into account his knowledge of the choice of chance will not be optimal.

Let us say at once that Player 2's optimal strategy is quite complicated (it will be discussed in Section 10.6). His strategy for the game Δ_∞ is simple: "Play left at each stage with probability 1/2." This is certainly *not* an optimal strategy for Γ_∞, because it can be answered by Player 1 going left if chance chooses left and right if chance chooses right. Such a strategy assures Player 1 an expected payoff of 1/2 at each stage, which is greater than the value of Γ_∞.

Let us now look more closely at optimal play in the game Δ_1 — this time from the point of view of Players 1_A and 1_B in Δ_1 (see Figure 10.4). Clearly, the expected payoff for Player 1_A is $u_A = 1/4$ and the expected payoff for Player 1_B is $u_B = 1/4$. (These are also the expected average payoffs if Δ_∞ is being played optimally by both players.) Note that $u_A > w_A = 0$ and $u_B > w_B = 0$. We may

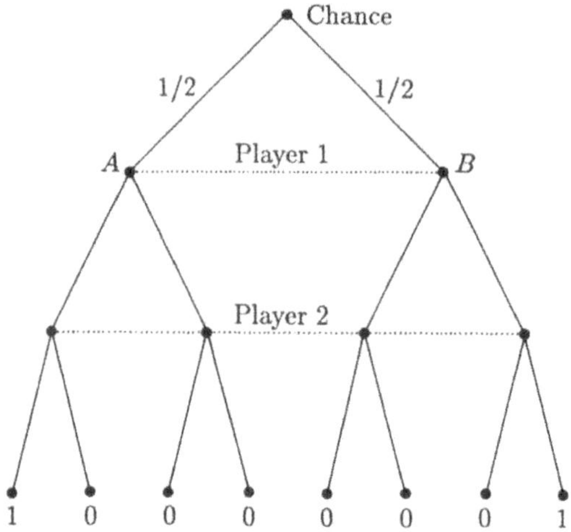

Figure 10.4: The game Δ_1.

summarize these remarks by saying that *if Player 1 commits himself in advance not to use his knowledge of the choice of chance, and if both players,* 1 *and* 2, *play optimally (in* Δ_∞*), then both* 1_A *and* 1_B *receive average expected payoffs that are greater than* w_A *and* w_B*, respectively.*

10.3. A Case When a Player Should Disclose His Information

In this section we use the notations of the previous section, except that here G_A and G_B denote the games

G_A (Player 1 rows, Player 2 columns):

Player 1 \ Player 2	L	R
L	-1	0
R	0	0

G_B (Player 1 rows, Player 2 columns):

Player 1 \ Player 2	L	R
L	0	0
R	0	-1

Here, again, $w_A = w_B = 0$.

In the last section it behooved Player 1 to act as though he did not know the choice of chance. Such an action is obviously wrong in this game. For in this case, Δ_1 is the game

		Player 2	
		L	R
Player 1	L	$-1/2$	0
	R	0	$-1/2$

whose value is $u_1 = -1/4$, and so equals the value of Δ_∞; whereas Player 1 can assure himself a higher expected average payoff in Γ_∞, namely 0, by playing always right if chance chooses left, and playing always left if chance chooses right. This way of playing is optimal for Player 1 because the entries in G_A and in G_B are nonpositive, so Player 1 assures himself the maximum amount by playing this way. Of course, if Player 1 plays optimally as above and Player 2 knows this, he will conclude, immediately after the first stage, what the choice of chance was, by observing Player 1's choice in the first stage. Thus, any way of playing while keeping this knowledge secret will not help Player 1.

Unlike in the previous section, here there is a simple optimal strategy for Player 2. It is: "play anything you wish." Any strategy chosen by Player 2 guarantees that he will not pay more than 0 to Player 1; and since Player 1 can guarantee 0, such a strategy is optimal.

Note that this means that in the original game Γ_∞, Player 1_A gets an average expected payoff $v_A = 0$, and similarly $v_B = 0$; hence $v_A = w_A$, $v_B = w_B$.

The absurdity of Player 1 playing in Γ_∞ as if he were playing optimally in Δ_∞ is also seen by comparing the values u_A and u_B with w_A and w_B. Indeed, if Players 1 and 2 play Δ_1 [or Δ_∞] optimally, then Player 1_A receives an expected [average] payoff of $-1/4$, and so does Player 1_B. Summarizing, we find that *if Player 1 commits himself in advance* not to use his knowledge of the choice of chance, and *if both Player 1 and Player 2 play optimally under this commitment (i.e., in Δ_∞), then both Player 1_A and Player 1_B receive expected payoffs that are less than w_A and w_B, respectively.* This commitment is silly because both Player 1_A and Player 1_B

can assure themselves a zero payoff at each stage by making use of their knowledge of the choice of chance; indeed, they can even allow Player 2 to deduce this.

10.4. A Case When a Player Should Partially Disclose His Information

In one of the previous examples we found that $u_A \geq w_A$ and $u_B \geq w_B$, and Player 1 could play optimally by not making any use of his knowledge of the choice of chance.[5]

In another example we found that $u_A < w_A$ and $u_B < w_B$, whence such behavior is certainly not optimal. In that case we found, however, that Player 1 should play as if he knew the choice of chance and, in fact, reveal the choice of chance immediately.

But these cases do not exhaust all the possibilities. It may happen that, say, $u_A > w_A$ and $u_B < w_B$. For example, let G_A and G_B be the games

G_A (rows: Player 1; columns: Player 2)

	L	R
L	1	0
R	0	2

G_B (rows: Player 1; columns: Player 2)

	L	R
L	−2	0
R	0	−1

(each being assigned probability 1/2). Their values are $w_A = 2/3$, $w_B = -2/3$. The tree of the game Δ_1 is drawn in Figure 10.5.

It is easy to verify that the only optimal strategy for Player 1 in this game is to play right and the only optimal strategy for Player 2 is to play left. (See the arrows in the figure.) The value of Δ_1 is $u = 0$.

Looking at the payoffs for Players 1_A and 1_B, we find $u_A = 0$ and $u_B = 0$ and therefore $u_A < w_A$ whereas $u_B > w_B$. The same numbers are, of course, obtained for Δ_∞ when the average expectation replaces the expected payoff.

[5]It will turn out that this in not always an optimal procedure even if $u_A \geq w_A$ and $u_B \geq w_B$.

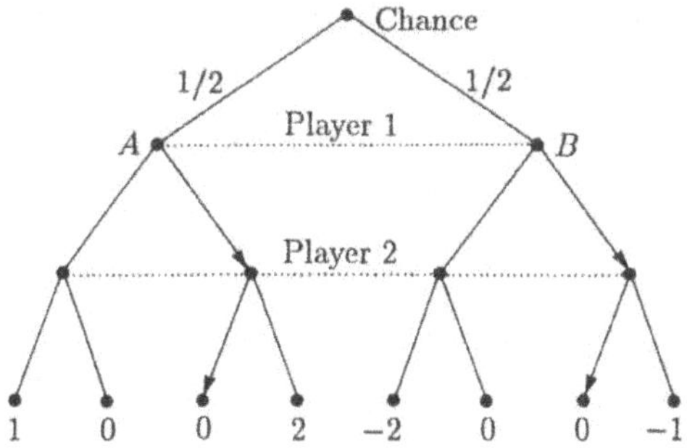

Figure 10.5: The tree of Δ_1.

Consider again the two alternatives for Player 1 playing Γ_∞, which were discussed at the beginning of this section:

i. Act as though he himself does not know the choice of chance (i.e., play as if he were playing optimally in Δ_∞).
ii. Reveal the choice of chance to Player 2 (i.e., play maximin against G_A or G_B at each stage in accordance with the choice of chance).

Suppose Player 1 has to choose among these alternatives after chance has made the decision (which corresponds to the real game we started with); then we are led to the following dilemma:

If chance chooses left, he would prefer (ii), because $u_A < w_A$. If chance chooses right, he would prefer (i), because $u_B > w_B$. But could he really do so effectively in Γ_∞? If he does, and attempts to conceal the choice of chance, Player 2 would recognize that Player 1 is taking the first alternative and would therefore infer that chance has chosen right.

These considerations indicate that there should be other alternatives that, so to say, partially disclose the choice of chance.

In fact, we shall see in this section that the value of Γ_∞ exists and is[6] 1/6 (instead of 0, which is the value of Δ_∞). Of course, the way in which Player 1 can guarantee himself this value is different from the above alternatives.

[6]The value itself can also be computed by the general theorem in Section 10.5.

Player 2 can quite easily make sure that in playing Γ_∞ he will not pay an expected average payoff greater than 1/6. All he has to do is choose left at each stage with probability 2/3, and right with probability 1/3. Knowing this, Player 1_B would certainly choose R at each stage, and no matter what Player 1_A will do, the expected payoff for the stage will be $\frac{1}{2}\cdot\frac{2}{3}+\frac{1}{2}\cdot(-\frac{1}{3})=\frac{1}{6}$.

Thus, the upper limit of the average payoffs for Player 1 is not greater than 1/6, regardless of his strategy choice.

We must still show that Player 1 can guarantee himself an average expected payoff that is at least 1/6. We shall actually find Player 1's optimal strategy: but first we have to compute the optimal strategy for Player 1 and the value of the 1-stage game $\Gamma_1(p)$, for an arbitrary p, $0 \le p \le 1$ (see Figure 10.6). The normal form of this game is

		Player 2	
		L	R
Player 1	LL	$3p-2$	0
	LR	p	$p-1$
	RL	$2p-2$	$2p$
	RR	0	$3p-1$

To compute Player 1's optimal strategy in $\Gamma_1(p)$, we consider several cases.

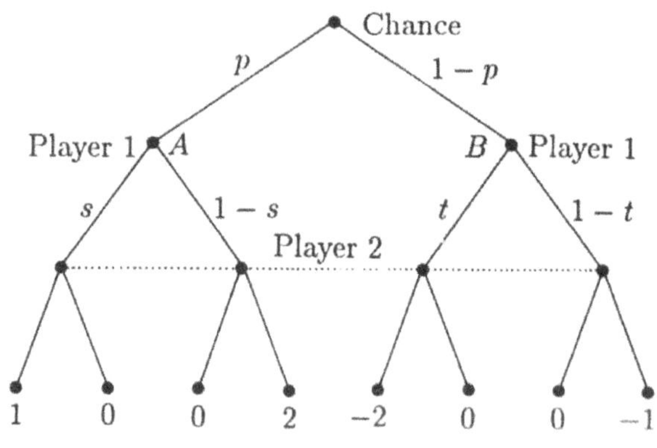

Figure 10.6: The game $\Gamma_1(p)$.

A. Suppose $0 < p < 1/3$; then only the last two strategies of Player 1 are active, and he should play the strategy RL with probability $\frac{1-3p}{3(1-p)}$ and RR with probability $\frac{2}{3(1-p)}$. The value of the game is

$$v_1(p) = \frac{-2(1-3p)}{3}, \quad 0 < p < 1/3. \tag{4.1}$$

In terms of the behavioral strategies described in Figure 10.6, Player 1 should play right if chance chose left, i.e., $s = 0$; and he should play left with probability $t = \frac{1-3p}{3(1-p)}$ if chance chose right.

B. Suppose $1/3 < p < 1$; then the strategies LR and RR are the only active strategies. Player 1's optimal strategies are LR with probability $(3p-1)/3p$ and RR with probability $1/3p$. The value of the game is

$$v_1(p) = \frac{(3p-1)}{3}, \quad 1/3 < p < 1. \tag{4.2}$$

In terms of the behavioral strategies described in Figure 10.6, Player 1 should play right if chance chose right, i.e., $t = 0$; and he should play left with probability $s = (3p-1)/3p$ if chance chose left.

C. Suppose $p = 1/3$. Player 1 should then play RR; i.e., he should choose right regardless of the choice of chance ($s = 0$, $t = 0$). The value of the game is

$$v_1(1/3) = 0. \tag{4.3}$$

D. Suppose $p = 0$ or $p = 1$. The values of the games in these cases are $-2/3$ and $2/3$, respectively.

Figure 10.7 shows the value of $\Gamma_1(p)$ as a function of p. Note that if $p = 1/2$, then $s = 1/3$, $t = 0$, and $v_1(1/2) = 1/6$. Observe also that $v_1(p) > 0$ for $p > 1/3$.

Let us turn now to the game $\Delta_\infty(1/2)$. We must exhibit a strategy σ for Player 1 that guarantees him 1/6 in Γ_∞. More precisely, a strategy σ such that

$$\liminf_{n\to\infty} \frac{1}{n} \sum_{i=1}^{n} \gamma_i(\sigma, \tau) \geq 1/6 \tag{4.4}$$

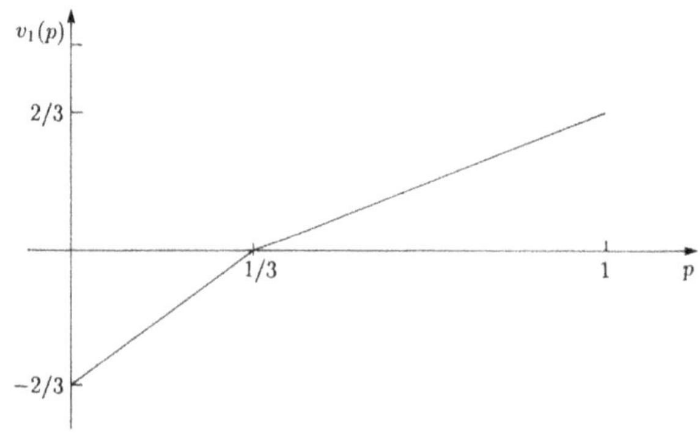

Figure 10.7: The values of $\Gamma_1(p)$.

for *every* strategy τ chosen by Player 2. Here, $\gamma_i(\sigma, \tau)$ is the expected payoff to Player 1 at the i-th stage of Γ_∞, obtained when Player 1 employs σ and Player 2 employs τ. Thus, for a given σ we must determine a strategy τ for Player 2 that minimizes the left side of (4.4).

The key word in the subsequent arguments is the word "every" above. Heuristically speaking, the following ideas will be helpful: a strategy that minimizes the left side of (4.4) is a strategy that Player 2 would have chosen *had he known* that Player 1 is employing σ. We therefore assume that Player 1 commits himself to σ and that Player 2 knows σ and knows that Player 1 is committed to σ. As far as Player 2 is concerned — Player 1 is now nothing but a chance mechanism playing σ and his object is to choose a pure strategy that minimizes (4.4).

As a result of playing σ, there is a choice of L or R at each stage with known probabilities, and by the rules of the game, this choice is revealed at the next stage. Knowing both the probability of the choice and the actual choice, Player 2 can compute the *conditional probabilities* that chance has chosen either vertex A (Game G_A) or vertex B (Game G_B). He can therefore make use of this computation for the next stage. All this amounts to is that Γ_∞ can be regarded as a sequence of 1-stage independent games $\Gamma_1(p_i)$ (see Figure 10.8), where s_i, t_i are the (known) probabilities that Player 1 chooses left at the i-th stage, at the vertices A and B, respectively, and p_i is

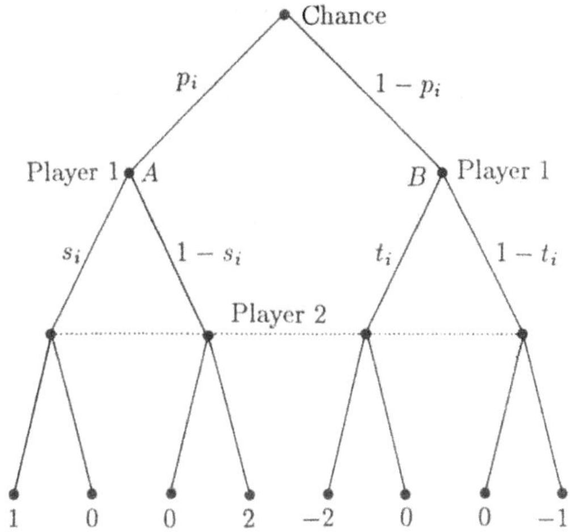

Figure 10.8: The game $\Gamma_1(p_i)$.

the conditional probability of chance choosing L given the choices of Player 1 prior to Stage i.

Intuitively, $\Gamma_1(p_i)$ is the game that Player 2 believes he is facing at stage i. It is therefore reasonable[7] to expect that the best reply for Player 2 is to choose L or R at each stage i so that the expected payoff for Player 1 in $\Gamma_1(p_i)$ is minimum, where s_i and t_i are the probabilities of Player 1 choosing L, induced by σ at stage i.

To be sure — the strategy that we have just described for Player 2 is *not* an optimal strategy for him in Γ_∞. It is only one of the best replies that Player 2 has *if* Player 1 is committed to σ. Hence it is a strategy for Player 2 that minimizes the left side of (4.4).

We are now in a position to describe an optimal strategy for Player 1. It is a behavioral strategy: *At each stage i, Player 1 should assume that he is facing the 1-stage game $\Gamma_1(p_i)$ and play optimally accordingly.*[8] Here, the p_i's are the conditional probabilities

[7]Formally, this follows from Theorem N4 in Postscript e to this chapter (p_2 in (N13) depends on σ but not on τ).

[8]Such a strategy is optimal in this specific game, but not necessarily in other games.

attributed to chance choosing A by a Player 2 who is assumed to know Player 1's strategy and previous moves.

Let us now compute this strategy stage by stage and then determine the expected payoff at each stage if Player 2 answers with his best reply.

At the beginning, Player 1 assumes that he is facing $\Gamma_1(1/2)$. He therefore plays $s_1 = 1/3$, $t_1 = 0$. As a result of this strategy two events might happen: Either he chooses left — the probability of this is 1/6 — or he chooses right. If he chooses left, the second stage is $\Gamma_1(1)$. If he chooses right, the second stage is $\Gamma_1(2/5)$. The situation is summarized in Figure 10.9.

If Player 1 chooses left, then Player 2 "realizes" that chance has decided upon game A. Thus, Player 1 will continue to play minimax

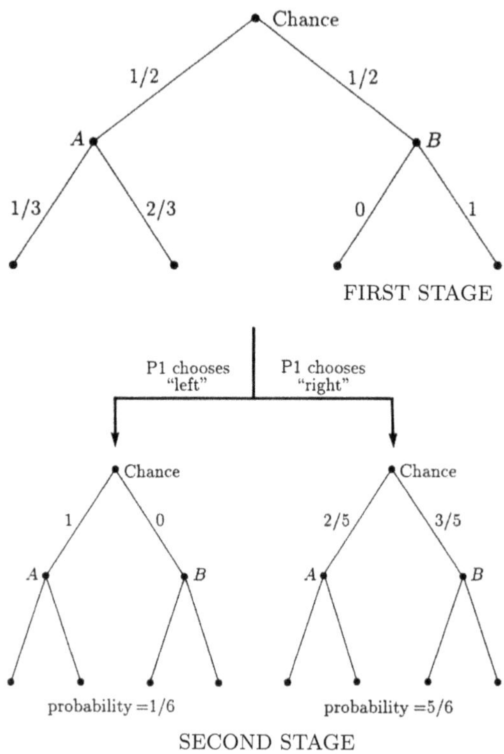

Figure 10.9: After the first stage.

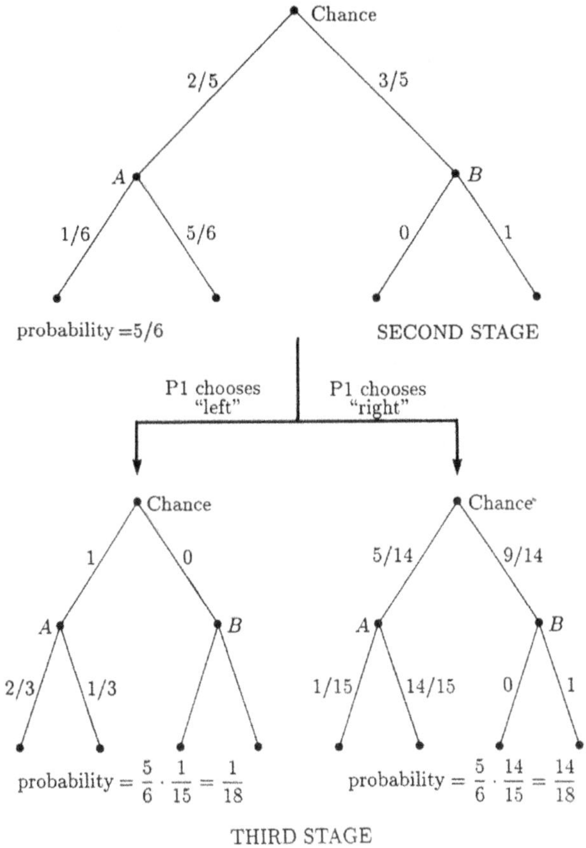

Figure 10.10: After the second stage.

against G_A *at each subsequent stage*; i.e., $s_i = 2/3$, $i = 2, 3, \ldots$. (His choice at vertex B is irrelevant because this happens with 0 probability.)

If Player 1 chooses right, then the conditional probability attributed by Player 2 to chance choosing A is 2/5. By our prescription, Player 1 should then play as if he were facing the 1-stage game $\Gamma_1(2/5)$; i.e., $s_2 = 1/6$, $t_2 = 0$. The picture is now as in Figure 10.10. Thus, if left was chosen, Player 1 will play $s_i = 2/3$ from stage 3 onwards. If right was chosen, Player 1 will play optimally for $\Gamma_1(5/14)$ in the third stage; i.e., $s_3 = 1/15$, $t_3 = 0$.

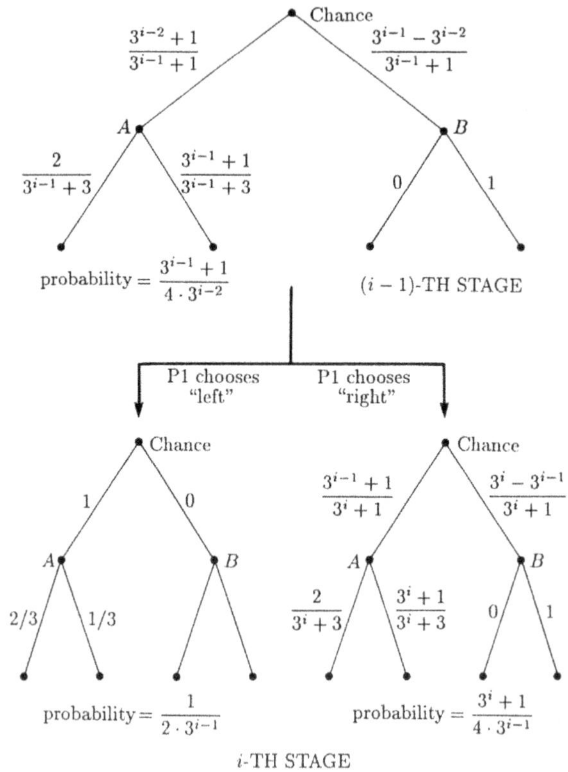

Figure 10.11: After the $(i-1)$-th stage.

It can be shown by induction that for $i = 2, 3, \ldots,$ the stages look as in Figure 10.11.

The case not described in this figure is that in which it was revealed in one of the previous stages $2, 3, \ldots, i-1$ that chance chose A. This occurs with probability

$$1 - \frac{3^{i-1}+1}{4\cdot 3^{i-2}} = \frac{3^{i-2}-1}{4\cdot 3^{i-2}}. \tag{4.5}$$

At each stage i, Player 1 plays an optimal strategy for a certain game $\Gamma_1(p_i)$ (where p_i depends on the sequence of choices made by Player 1 at the previous stages). As indicated previously, a best reply for Player 2 is to play optimally for the *same* game $\Gamma_1(p_i)$. Therefore, at stage i Player 1 will obtain an expected payoff of $v_1(p_i)$. It follows from the p_i's being conditional probabilities determined by his own strategy, that Player 1's expected payoff for the i-th stage of Γ_∞ is

$$\begin{aligned}
&\text{prob}(p_i = 1) \cdot v_1(1) + \text{prob}\left(p_i = \frac{3^{i-1}+1}{3^i+1}\right) \cdot v_1\left(\frac{3^{i-1}+1}{3^i+1}\right) \\
&= \left(\frac{3^{i-2}-1}{4\cdot 3^{i-2}} + \frac{1}{2\cdot 3^{i-1}}\right)\cdot\frac{2}{3} + \frac{3^i+1}{4\cdot 3^{i-1}}\cdot v_1\left(\frac{3^{i-1}+1}{3^i+1}\right) \\
&= \frac{1}{6} - \frac{1}{2\cdot 3^i} + \frac{3^i+1}{4\cdot 3^{i-1}}\cdot v_1\left(\frac{3^{i-1}+1}{3^i+1}\right) \\
&\geq \frac{1}{6} - \frac{1}{2\cdot 3^i}.
\end{aligned} \tag{4.6}$$

The last inequality follows from $\frac{3^{i-1}+1}{3^i+1} > \frac{1}{3}$; therefore $v_1\left(\frac{3^{i-1}+1}{3^i+1}\right)$ is positive (see Figure 10.7 and the remarks referring to it in the text). Thus, at each stage, Player 1 guarantees by his strategy to get at least $1/6 - (2\cdot 3^{\mathrm{i}})^{-1}$; hence, he can guarantee in Γ_∞ a lower limit of the average expected payoff that is at least $1/6$. When we discussed the optimal strategy of Player 2 in Γ_∞, we saw that Player 2 can guarantee that the upper limit of the average expected payoff will not exceed $1/6$; therefore, the infinite game has a value of $1/6$, and the above strategy is optimal for Player 1.

Note that the probability that chance's choice will eventually be disclosed is $\lim_{i\to\infty}(3^{i-2}-1)/4\cdot 3^{i-2} = {}^1\!/_4$. For this reason we say that in this situation, Player 1 *partially discloses the choice of chance.* By substituting (4.2) in (4.6), it can be shown that the above strategy assures Player 1 an expected payoff of at least $1/6$, which is precisely $v_1(1/2)$, *at each stage.* Thus, the strategy is optimal even for the n-stage game Γ_n.

10.5. The Main Theorem

In the preceding sections we showed by means of examples that in some cases an optimal strategy for Player 1 in a game Γ_∞ is to refrain from making use of his knowledge of chance's choice; i.e., to play as if he were in the game Δ_∞. In other cases, his optimal strategy makes use of his knowledge, which may result in fully or partially revealing chance's choice to Player 2 (who is supposed to know which strategy Player 1 is using).

The first question that comes to mind is: When should Player 1 play like in Δ_∞ and when not?

Obviously, Player 1 should make use of his knowledge of chance's choice if and only if he can guarantee an average expected payoff that is higher than the value of Δ_1. This brings us to the question of the value of Γ_∞ (if it exists), which is an important question in itself. In this section we answer this seemingly more difficult question, from which the answer to the first question will become apparent.

Let $\Gamma_\infty(p)$ be a general (infinite) game based on two 2-person constant sum games G_A and G_B, as sketched in Figure 10.12, where the rules are as explained in the previous sections. In particular, Player 1 knows the choice of chance, whereas Player 2 does not.

Let $\Delta_1(p)$ be the 1-stage game sketched in Figure 10.13. Let $u(p)$ be the value of $\Delta_1(p)$ (which is also the value of the infinite game $\Delta_\infty(p)$). It is a continuous function of p. Clearly $u(0) = w_B$ and $u(1) = w_A$, where w_A and w_B are the values of G_A and G_B, respectively.

In general, u is not a concave function, and therefore it may be different from the function cav u defined by

$$\text{cav}\, u(p) := \min \tilde{u}(p), \quad 0 \leq p \leq 1, \tag{5.1}$$

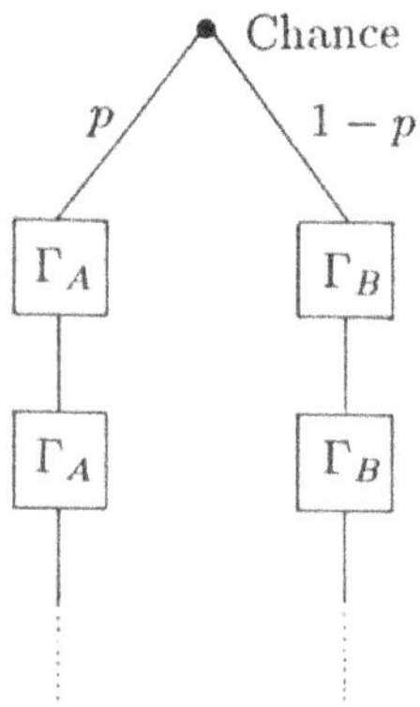

Figure 10.12: The game $\Gamma_\infty(p)$.

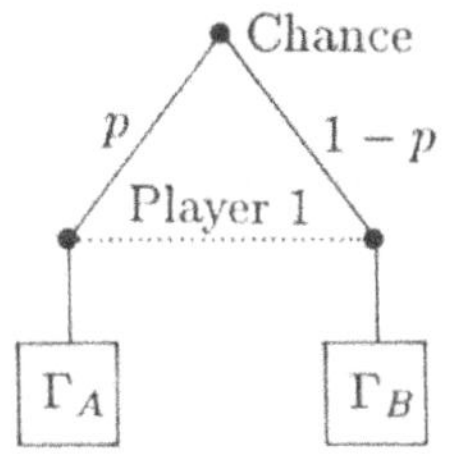

Figure 10.13: The game $\Delta_1(p)$.

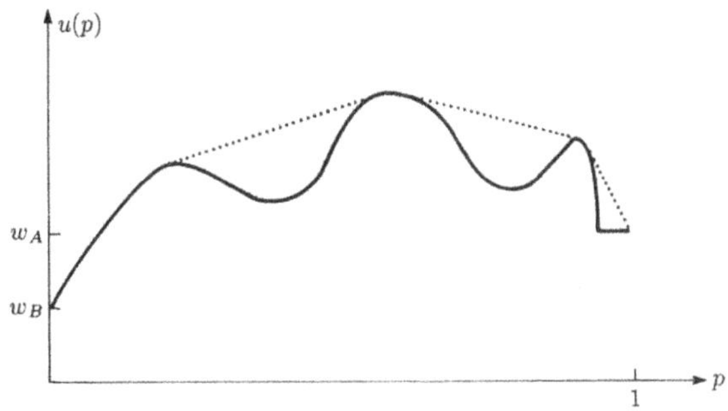

Figure 10.14: The concavification of $u(p)$.

where the minimum is taken over all *concave* functions $\tilde{u}$ such that $\tilde{u}(p) \geq u(p)$ for all p. The function cav u is itself concave; it will be called *the concavification* of u. It is indicated in Figure 10.14 by a broken line for all the values p where cav $u(p) \neq u(p)$.

Theorem 5.1. *The value of the game* $\Gamma_\infty(p)$ *exists, and is equal to* cav $u(p)$, *where* cav u *is the concavification of* u.

We outline the proof in the remainder of the section, for simplicity taking only G_A and G_B that are 2×2 matrix games. Player 1's (behavioral) strategy is then an infinite sequence of pairs (s_1, t_1), $(s_2, t_2), \ldots$, where s_i, t_i are the probabilities at stage i of Player 1 choosing "left" if chance chose "left" or "right", respectively.[9]

[9]By the theorem of Kuhn [1953], generalized to infinite games [Aumann, 1964], we can limit ourselves to behavioral strategies, because ours is a game of perfect recall.

Similarly, Player 2's strategy is a sequence $\lambda_1, \lambda_2, \ldots,$ where λ_i is the probability of "left" at the i-th stage. The parameters s_i, t_i, λ_i may depend on the actual choices made by both players at the previous stages.

Lemma 5.2. *Let* $0 \le p^L < p < p^R \le 1$, *or* $0 \le p^R < p < p^L \le 1$; *then there exist probabilities* s_1, t_1 *for the first stage, such that*

$$\text{prob}\left\{chance\ chose\ \text{“}left\text{”} \,\middle|\, \begin{matrix} player\ 1\ went\ left \\ at\ the\ first\ stage \end{matrix}\right\} = p^L, \tag{5.2}$$

$$\text{prob}\left\{chance\ chose\ \text{“}left\text{”} \,\middle|\, \begin{matrix} player\ 1\ went\ right \\ at\ the\ first\ stage \end{matrix}\right\} = p^R. \tag{5.3}$$

Proof. The equations (5.2) and (5.3) are, in fact,

$$p^L = \frac{s_1 p}{s_1 p + t_1(1-p)}, \quad p^R = \frac{(1-s_1)p}{(1-s_1)p + (1-t_1)(1-p)}, \tag{5.4}$$

whose solution is:

$$s_1 = \frac{p^L(p^R - p)}{p(p^R - p^L)}, \quad t_1 = \frac{(1-p^L)(p^R - p)}{(1-p)(p^R - p^L)}. \tag{5.5}$$

Clearly $0 \le s_1 < 1$ and $0 \le t_1 \le 1$.

Lemma 5.3. *Player 1 can guarantee in* $\Gamma_\infty(p)$ *average expected payoffs whose lower limit is at least* cav $u(p)$.

Proof. We shall exhibit a strategy for Player 1 such that even if Player 2 knew it and knew that Player 1 is committed to it, his best reply would still yield Player 1 an average expected payoff whose lower limit is not less than cav $u(p)$.

A. If p is such that cav $u(p) = u(p)$, then Player 1 acts as if he does not know the choice of chance; i.e., plays as if he is playing Δ_1. In other words, he plays at each stage as if he were playing optimally in Δ_1. He then guarantees an average expected value whose lower limit is at least cav $u(p)$.

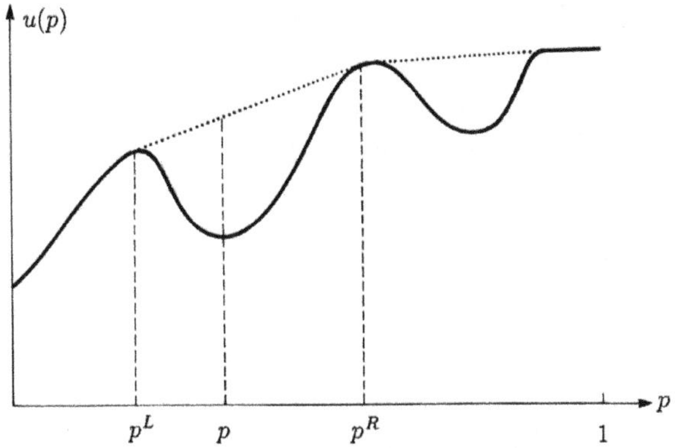

Figure 10.15: $u(p^L) = \text{cav}\, u(p^L)$ and $u(p^R) = \text{cav}\, u(p^R)$.

B. If p is such that cav $u(p) > u(p)$, then there exist p^L and p^R with $0 \leq p^L < p < p^R \leq 1$, such that $u(p^L) = $ cav $u(p^L), u(p^R) =$ cav $u(p^R)$ and (p^L, cav $u(p^L)$), (p, cav $u(p)$), and (p^R, cav $u(p^R)$) are on a straight line (see Figure 10.15).

A strategy for Player 1 is as follows: Choose s_1 and t_1 satisfying (5.5). If left was played at the first stage, continue as in $\Delta_\infty(p^L)$; i.e., play an optimal strategy for $\Delta_1(p^L)$ at each stage i, $i \geq 2$.

If right was played at the first stage, continue as in $\Delta_\infty(p^R)$; i.e., play an optimal strategy for $\Delta_1(p^R)$ at each stage $i, i \geq 2$.

Player 2, who is assumed to know that Player 1 is committed to this strategy, would decide after the choices of the first stage are revealed to play as in $\Delta_\infty(p^L)$, if at the first stage Player 1 chose left; and as in $\Delta_\infty(p^R)$, if Player 1 chose right. His best reply would still guarantee Player 1 average expected payoffs with lower limit either $u(p^L)$ or $u(p^R)$, according to his choice in the first stage.[10]

The probability that Player 1 chooses left at the first stage is $ps_1 + (1 - p)t_1 =: \alpha$. The probability that he chooses right is $p(1 - s_1) + (1 - p)(1 - t_1) = 1 - \alpha$. Observe that

$$\alpha p^L + (1 - \alpha)p^R = p. \tag{5.6}$$

[10]Whatever Player 1 has gained in the first stage has no effect on the average expected payoffs when Γ_∞ is played.

In other words, p is the center of gravity of the weights α and $1 - \alpha$ attached to p^L and p^R, respectively. Consequently, $(p,\ \text{cav}\ u(p))$ is the center of gravity of the same weights attached to the points $(p^L,\ u(p^L))$ and $(p^R,\ u(p^R))$, because these points are on a straight line. It follows that

$$\begin{aligned}\alpha u(p^L) + (1-\alpha)u(p^R) &= \alpha\,\text{cav}\,u(p^L) + (1-\alpha)\,\text{cav}\,u(p^R)\\ &= \text{cav}\,u(p). \qquad (5.7)\end{aligned}$$

But the left side of (5.7) is the lower limit of the average expected payoffs for Player 1. We have therefore shown that this lower limit is equal to cav $u(p)$. This concludes the proof.

The relation (5.6) can be interpreted as follows: The (unconditional) expectation of the conditional probability at the second stage that chance has chosen "left" is equal to the original probability that chance has chosen "left". This, of course, can be generalized to an arbitrary stage.

Thus, let p_n be the conditional probability that chance has chosen left, given the choices of Player 1 at stages $1, 2, \ldots, n-1$, and his announced strategy $\{(s_i,\ t_i)\}$. Then p_n is a random variable taking up to 2^{n-1} possible values, and

$$\text{Exp}(p_n) = p. \qquad (5.8)$$

(Here, $p_1 = p$.) We must still show that Player 2 can guarantee that he will not pay an average expected payoff whose expected limit is greater than cav $u(p)$.

We first give this part of the proof for the case treated in Section 10.2, where G_A and G_B are the games

Player 1 \ Player 2	L	R
L	1	0
R	0	0

G_A

Player 1 \ Player 2	L	R
L	0	0
R	0	1

G_B

and furthermore $p = 1/2$. We shall later indicate what modifications should be made in the proof of the general case, which is technically more complicated but uses the same ideas.

Before giving this part of the proof technically, we describe the idea. It is based on the minimax theorem. If to each strategy of Player 1, Player 2 can respond so as to guarantee paying at most 1/4 on the average, then, in a finite game, the minimax theorem tells us that he has a strategy that guarantees at most 1/4. Therefore, we may assume that Player 1 must announce his behavior strategy $\{(s_i, t_i)\}$ before the start of the game. If s_i and t_i differ from each other by "a significant amount" and "sufficiently often", then Player 2 will be able to deduce chance's choice with high probability, and the resulting payoff will be close to 0. Otherwise, s_i and t_i must eventually become and remain "almost equal." When this happens, Player 2 can play in the resulting game — in which the probabilities are now the conditionals p_i — as if Player 1 did not know chance's choice either. If Player 2 then plays optimally for the one-shot case, he will get a payoff of $p_i(1 - p_i)$, which is always at most 1/4. We now turn to the technical details.

Step 1. We first compute $u(p)$ and $v_1(p)$ for $0 \leq p \leq 1$. These are the values of the games whose normal forms are

	L	R
L	p	0
R	0	$1-p$

	L	R
LL	p	0
LR	p	$1-p$
RL	0	0
RR	0	$1-p$

Here, $u(p) = p(1-p) = \text{cav}\, u(p)$, and $v_1(p) = \min(p, 1-p)$ (see Figure 10.16). Note that $\text{cav}\, u(p)$ and $v_1(p)$ are continuous functions of p.

Step 2. We now assume that in the n-stage game $\Gamma_n(1/2)$, Player 1 announces and commits himself to a (not necessarily optimal) strategy $\{(s_i, t_i)\}$, $i = 1, 2, \ldots, n$, and that Player 2 knows this. A possible reply for Player 2 in $\Gamma_n(1/2)$ would be to answer at each

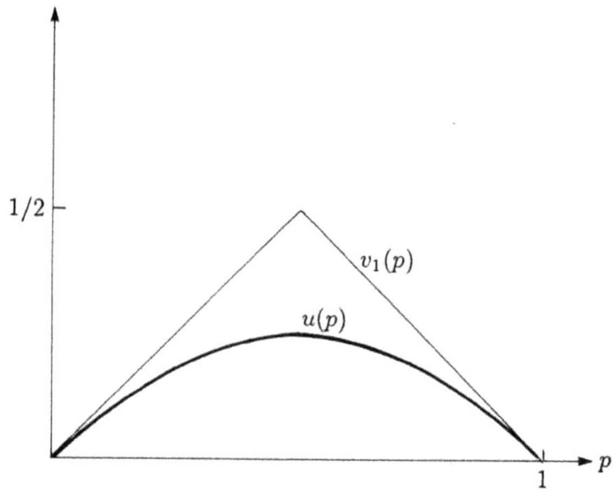

Figure 10.16: The graphs of $u(p)$ and $v_1(p)$.

stage i as best he can, as if it were a 1-stage game based on the conditional probability p_i that chance chose "left". The payoff to Player 1 at stage i (as perceived by Player 2) will then be:

$$\begin{aligned} g_i &= \min\{p_i s_i, (1-p_i)(1-t_i)\} \\ &= \min\{p_i s_i, (1-p_i)(1-s_i) + (1-p_i)(s_i - t_i)\} \\ &\le \min\{p_i s_i, (1-p_i)(1-s_i)\} + |s_i - t_i| \qquad (5.9) \\ &\le u(p_i) + |s_i - t_i| \le 1/4 + |s_i - t_i|, \quad i = 1, 2, \ldots, n. \end{aligned}$$

(The second inequality follows from the fact that if Player 1 plays any $(s_i, 1-s_i)$ in $\Delta_1(p_i)$, he cannot guarantee more than $u(p_i)$.)

Step 3. It follows from (5.5), adapted to stage i, that

$$|s_i - t_i| = \left| \frac{(p_{i+1}^R - p_i)(p_{i+1}^L - p_i)}{p_i(1-p_i)(p_{i+1}^R - p_{i+1}^L)} \right|, \quad i = 1, 2, \ldots, n, \qquad (5.10)$$

where[11] p_{i+1}^R, p_{i+1}^L are the conditional probabilities that chance chose left, given p_i at the previous stage, in case Player 1's choice at stage i

[11] p_{n+1}^R and p_{n+1}^L are meaningful even if the game terminates after n stages.

is right or left, respectively. Consequently, in any case,

$$|s_i - t_i| \le K|p_{i+1} - p_i|, \quad \text{if, say, } \frac{1}{4} \le p_i \le \frac{3}{4}, \tag{5.11}$$

where K is a positive constant, and p_{i+1} is either p^R_{i+1} or p^L_{i+1}. Indeed, p_i is a weighted average of p^R_{i+1} and p^L_{i+1}, and it follows that $\left|p^R_{i+1} - p^L_{i+1}\right| \ge \left|p^R_{i+1} - p_i\right|$ and $\left|p^R_{i+1} - p^L_{i+1}\right| \ge \left|p^L_{i+1} - p_i\right|$.

Step 4. It follows from the expression for $v_1(p)$ in Step 1 that for each i, $i = 1, 2, \ldots, n$,

$$g_i \le \frac{1}{4} \text{ if either } p_i \le \frac{1}{4} \text{ or } p_i \ge \frac{3}{4}. \tag{5.12}$$

Consequently, by (5.11),

$$g_i \le \frac{1}{4} + K|p_{i+1} - p_i|. \tag{5.13}$$

Step 5. We are interested in Exp $\sum_{i=1}^{n} g_i$ and so must estimate the expression

$$\text{Exp} \sum_{i=1}^{n} |p_{i+1} - p_i|\,.$$

Fortunately, the conditional probabilities p_i satisfy

$$\text{Exp}(p_i \,|\, p_1, p_2, \ldots, p_{i-1}) = p_{i-1}, \quad i = 1, 2, \ldots, n. \tag{5.14}$$

Indeed, (5.14) follows from the fact that given $p_1, p_2, \ldots, p_{i-1}$, it is possible to compute Exp(p_i) by an analogue of (5.6), using p_{i-1} only. A sequence of random variables p_i satisfying (5.14) is known as a *martingale*; and for martingales it may be shown (see, e.g., Doob [1953]) that

$$\text{Exp} \sum_{i=1}^{n} (p_{i+1} - p_i)^2 = \text{Exp}(p_{n+1} - p_1)^2. \tag{5.15}$$

Since, moreover, p_{n+1} and p_1 are probabilities, it follows that

$$\operatorname{Exp}\sum_{i=1}^{n}(p_{i+1}-p_i)^2 \leq 1. \tag{5.16}$$

Step 6. It follows from (5.13), (5.16) and the Cauchy-Schwarz inequality that

$$\begin{aligned}
\operatorname{Exp}\sum_{i=1}^{n} g_i &\leq \frac{n}{4} + K\operatorname{Exp}\sum_{i=1}^{n}|p_{i+1}-p_i|\cdot 1\\
&\leq \frac{n}{4} + K\operatorname{Exp}\sqrt{\sum (p_{i+1}-p_i)^2\cdot n}\\
&\leq \frac{n}{4} + K\sqrt{n}\operatorname{Exp}\sqrt{\sum (p_{i+1}-p_i)^2+1}\\
&\leq \frac{n}{4} + K\sqrt{n}\operatorname{Exp}\left(\sum (p_{i+1}-p_i)^2+1\right)\\
&\leq \frac{n}{2} + 2K\sqrt{n};
\end{aligned}$$

consequently,

$$\operatorname{Exp}\frac{1}{n}\sum_{i=1}^{n} g_i \leq \frac{1}{4} + \frac{2K}{\sqrt{n}} =: b_n, \tag{5.17}$$

and this expression tends to $1/4$ when n tends to infinity.

Step 7. Define the payoff in $\Gamma_n(1/2)$ to be the average payoff for the n stages in the game. By the minimax theorem and (5.17), the value v_n of $\Gamma_n(1/2)$ is not greater than b_n. Let σ_n be an optimal strategy for Player 2 in $\Gamma_n(1/2)$. We now describe an optimal strategy for Player 2 in $\Gamma_\infty(1/2)$: Play σ_1 in the first stage. Play σ_2 in the next two stages. Play σ_3 in the next three stages, etc. Let m be any integer; we shall compute the average expected payoff for m stages. Put $m = 1+2+3+\cdots+n+k$, where $0 \leq k < n+1$. The average expected payoff for m stages is at most

$$q_m := \frac{1}{m}\left(v_1 + 2v_2 + \cdots + nv_n + \frac{k}{2}\right)$$
$$\leq \frac{b_1 + 2b_2 + \cdots + nb_n}{1 + 2 + \cdots + n} + \frac{1}{n}, \qquad (5.18)$$

where, for the last k stages we took the maximum expected payoff in each game; namely, $1/2$. The last term in (5.18) tends to zero as n tends to infinity. The first term is a Cesaro mean of $n(n+1)/2$ terms of the sequence $b_1, b_2, b_2, b_3, b_3, b_3, \ldots$. By (5.17), this sequence tends to $1/4$ as $n \to \infty$; hence $\limsup q_m \leq 1/4$.

We have exhibited a strategy for Player 2 that guarantees that the upper limit of the average expected payoff is at most $1/4$. This and Lemma 5.3 prove Theorem 5.1 for the case of $\Gamma_\infty(1/2)$ treated here.

The proof in the general case is quite similar and we only outline the necessary modifications. In this case, p is an arbitrary number between 0 and 1, and G_A and G_B are arbitrary $\mu \times \nu$ matrix games, $\mu, \nu \geq 2$, whose values are w_A and w_B, respectively.

From Step 1 we use only the fact that $\operatorname{cav} u(p)$ and $v_1(p)$ are continuous functions of p that satisfy $v_1(0) = w_B = u(0) = \operatorname{cav} u(0)$ and $v_1(1) = w_A = u(1) = \operatorname{cav} u(1)$. In Step 2 an analogue of (5.9), which is easily derived, is

$$g_i \leq u(p_i) + C\|s_i - t_i\| \leq \operatorname{cav} u(p_i) + C\|s_i - t_i\|, \qquad (5.19)$$

where $s_i = (s_i^1, s_i^2, \ldots, s_i^\mu)$ and $t_i = (t_i^1, t_i^2, \ldots, t_i^\mu)$ are the i-th stage strategies of Player 1_A and Player 1_B, respectively. Now $\|s_i - t_i\|$ is the distance between the vectors s_i and t_i in, say, the maximum norm, and C is μ times a constant that is larger than the absolute value of each entry in G_A and G_B. Furthermore, we note that for each ε there is a δ such that if $p_i \leq \delta$ or $(1 - p_i) \leq \delta$, then

$$g_i \leq \operatorname{cav} u(p_i) + \varepsilon. \qquad (5.20)$$

This can be seen as follows: In any case we have $g_i < v_1\ (p_i)$. If p_i is small then from the continuity of v_1 in p it follows that $v_1(p_i)$ is close to $v_1(0) = \operatorname{cav} u(0)$, and from the continuity of cav u in p it follows that cav $u(0)$ is then close to cav $u(p_i)$. The argument is similar when p_i is close to 1.

One substantive change is made in Step 3: If p^r_{i+1} is the conditional probability of chance choosing left, given that Player 1 has chosen the r-th row at stage i (and given the conditional probability p_i), then

$$p^r_{i+1} = \frac{p_i s^r_i}{p_i s^r_i + (1-p_i)t^r_i}. \tag{5.21}$$

Hence,

$$(p_i s^r_i + (1-p_i)t^r_i)^2 (p^r_{i+1} - p_i)^2 = (1-p_i)^2 p^2_i (s^r_i - t^r_i)^2. \tag{5.22}$$

If $\delta < p_i < 1-\delta$, then since $0 \le p_i s^r_i + (1-p_i)t^r_i \le 1$, by setting $K = C/(\delta(1-\delta))$ we obtain

$$\begin{aligned}
(s^r_i - t^r_i)^2 &\le \frac{K^2}{C^2}(p^r_{i+1} - p_i)^2 (p_i s^r_i + (1-p_i)t^r_i)^2 \\
&\le \frac{K^2}{C^2}(p^r_{i+1} - p_i)^2 (p_i s^r_i + (1-p_i)t^r_i) \\
&\le \frac{K^2}{C^2}\sum_r (p^r_{i+1} - p_i)^2 (p_i s^r_i + (1-p_i)t^r_i) \\
&= \frac{K^2}{C^2}\operatorname{Exp}((p_{i+1} - p_i)^2 | p_i).
\end{aligned} \tag{5.23}$$

Since this is true for any r, it follows that

$$\|s_i - t_i\|^2 \le \frac{K^2}{C^2}\operatorname{Exp}(p_{i+1} - p_i)^2 | p_i.$$

Hence, from (5.19), we obtain

$$\begin{aligned} g_i &\leq \operatorname{cav} u(p_i) + C\|s_i - t_i\| \\ &\leq \operatorname{cav} u(p_i) + K\sqrt{\operatorname{Exp}(p_{i+1} - p_i)^2 | p_i} \\ &\leq \operatorname{cav} u(p_i) + \varepsilon + K\sqrt{\operatorname{Exp}(p_{i+1} - p_i)^2 | p_i}, \end{aligned}$$

when $\delta < p_i < 1 - \delta$, and this is true also when $p_i \leq \delta$ or $p_i \geq 1 - \delta$, because of (5.20). Hence, from the concavity of cav u and of $\sqrt{}$, and from Exp $p_i = p$, we obtain

$$\operatorname{Exp} g_i \leq \operatorname{cav} u(p) + \varepsilon + K\sqrt{\operatorname{Exp}(p_{i+1} - p_i)^2}.$$

Using the Cauchy-Schwarz inequality and (5.16), we obtain

$$\begin{aligned} \operatorname{Exp} \sum_{i=1}^{n} g_i &\leq n(\operatorname{cav} u(p) + \varepsilon) + K\sqrt{n}\sqrt{\sum_{i=1}^{n} \operatorname{Exp}(p_{i+1} - p_i)^2} \\ &\leq n(\operatorname{cav} u(p) + \varepsilon) + K\sqrt{n}. \end{aligned}$$

Thus,

$$v_n \leq \operatorname{Exp} \frac{1}{n} \sum_{i=1}^{n} g_i \leq \operatorname{cav} u(p) + \varepsilon + \frac{K}{\sqrt{n}}.$$

Finally, letting $n \to \infty$, we deduce that $\lim_{n\to\infty} v_n \leq \operatorname{cav} u(p) + \varepsilon$. Since this is true for arbitrary ε, we have $\lim_{n\to\infty} v_n \leq \operatorname{cav} u(p)$; and the rest of the proof is as before.

We remark that by a slight modification we can prove, as before, that

$$v_n(p) = \operatorname{cav} u(p) + \mathrm{O}\left(\frac{1}{\sqrt{n}}\right). \tag{5.24}$$

Theorem 5.1 remains true also in the more general situation in which chance chooses any of κ alternatives, each leading to a possibly different game. The proof follows the same pattern as when $\kappa = 2$, except that at one stage an induction on κ must be employed.

10.6. The Optimal Strategy of Player 2

The reader will have observed that Theorem 5.1 was proved without presenting an optimal strategy for Player 2. We shall now present such a strategy for the game treated in Section 10.2.

Rather than representing the situation as a tree, let us consider a somewhat different approach. We start with a 1-stage game.

Since Player 2 does not know chance's choice, he may regard his outcome[12] as a *pair* of numbers (x, y), where x represents the payoff if he plays against Player 1_A, and y represents the payoff if he plays against Player 1_B. Only at the end does chance reveal its choice. The situation for Player 2 can therefore be described by a bimatrix, an outcome space X and the original p, namely, $p = 1/2$. In our case, we obtain the following bimatrix and the outcome space of Figure 10.17.

		Player 2: L	Player 2: R
Player 1	L	$(1, 0)$	$(0, 0)$
	R	$(0, 0)$	$(0, 1)$

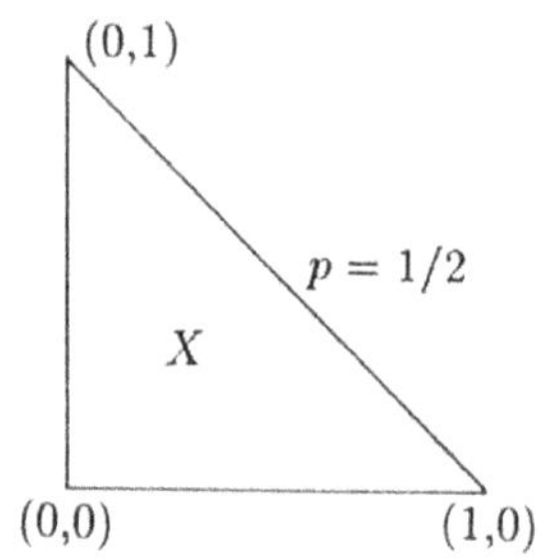

Figure 10.17: The outcome space.

[12] "Outcome" in this section is to be distinguished from "expected payoff."

If $\Delta_1(1/2)$ is being considered, we may say that Player 1, too, is informed of the choice of chance only at the end of the game. Therefore, if $(x,\ y)$ is the outcome, the expected payoff for Player 1 is $(x+y)/2$. The situation is somewhat different if $\Gamma_1(1/2)$ is being played, because in this case Player 1 knows the choice of chance at the beginning, and at the same time he has to take into account the game as viewed by Player 2. Suppose, for example, that Player 2 announces a mixed strategy $(q,\ 1-q)$; then Player 1 can answer either L or R, which lead to two points $(x_1,\ y_1)$ and $(x_2,\ y_2)$ on the axes in the outcome space. He can also answer by a mixed strategy, in which case he can achieve any point on the straight line segment l joining these points (see Figure 10.18).

If he wishes, his answer if he is Player 1_A may differ from his answer if he is Player 1_B. Knowing this, Player 2 must regard his actual expected payoff as

$$\max\left\{\frac{x+\eta}{2} : (x,y) \in l \text{ and } (\xi,\eta) \in l\right\}. \tag{6.1}$$

Thus, the expected payoff that Player 2 can guarantee is determined by considering *pairs of points* from a given set. A similar situation holds if an n-stage game is being considered, in which case one adds the vectors obtained as outcomes in each stage. The expected payoff for $\Delta_n(1/2)$ may be obtained by maximizing $(x_i + y_i)/2$ at each stage over a certain straight line segment l_i. The expected payoff for

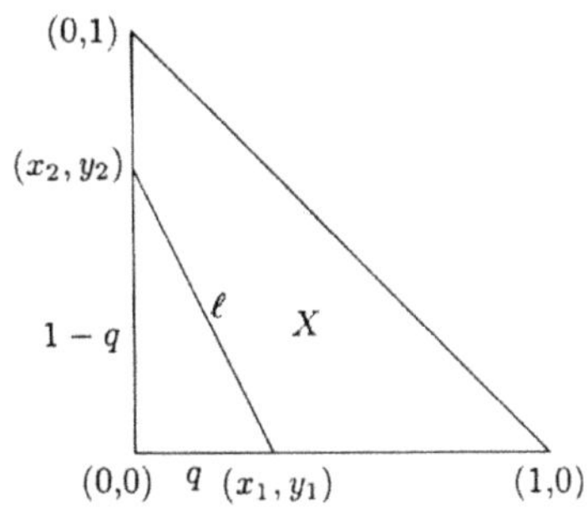

Figure 10.18: Payoffs against $(q, 1-q)$.

$\Gamma_n(1/2)$ may be obtained by maximizing at each stage an expression of the form (6.1) over all the l_i. In the case of an infinite game, one considers a *sequence* of *averages* of the two-dimensional vectors, i.e., their centers of gravity for $1, 2, 3, 4, \ldots$ stages, and one studies their limit points.

Games with vector outcomes were considered by D. Blackwell [1956], where he gave a criterion for a convex set S in X to be *approachable* by, say, Player 2 when the infinite game is played. By this is meant that Player 2 has a strategy such that with probability 1, the distance between a_i and S approaches 0 as $i \to \infty$, *regardless* of the strategy chosen by Player 1, where a_i is the average of the outcomes up to Stage i. It follows from his criterion that in our game $\Gamma_\infty(1/2)$, the square S of Figure 10.19 is an approachable set. This square is of size $1/4 \times 1/4$; two of its sides are on the axes and meet at the origin. Fortunately,

$$\max\left\{\frac{(x+\eta)}{2}: (x,y) \in S, \quad (\xi,\eta) \in S\right\} \leq \frac{1}{4}, \tag{6.2}$$

and since $1/4$ is the value of $\Gamma_\infty(1/2)$, a strategy for Player 2 that would lead the outcomes to a zero distance from S with probability 1 is optimal.

D. Blackwell [1956] describes a class of such strategies, one of which can be stated verbally as follows:

At each stage $i+1$, Player 2 observes his and Player 1's choices at the previous stages. He then computes the average outcome $(\bar{\xi}_i, \bar{\eta}_i)$

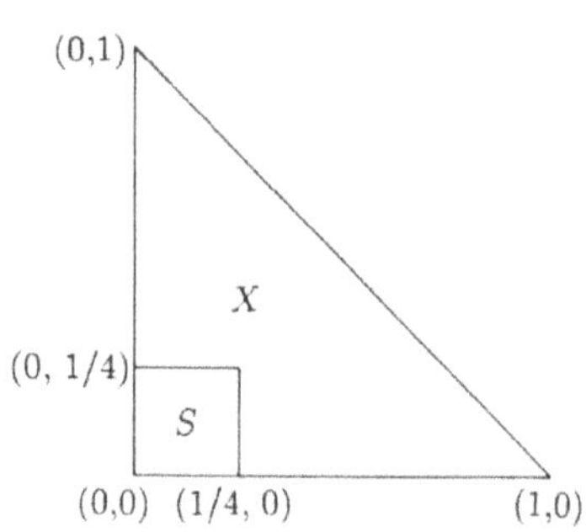

Figure 10.19: An approachable set.

for the first i stages. His intention is to see to it that both $\bar{\xi}_i$ and $\bar{\eta}_i$ will have upper limits that are not greater than 1/4. This he does as follows:

i. If $\bar{\xi}_i \leq 1/4$ and $\bar{\eta}_i \leq 1/4$, he is satisfied, and can play in any way he wishes at the $(i+1)$-th stage.
ii. If $\bar{\xi}_i \leq 1/4$ and $\bar{\eta}_i > 1/4$, then Player 1_A is O.K., but Player 1_B is not. Player 2 therefore plays against Player 1_B at the $(i+1)$-th stage; i.e., he plays L.
iii. Similarly, if $\bar{\xi}_i > 1/4$ and $\bar{\eta}_i \leq 1/4$, Player 2 plays R at the $(i+1)$-th stage.
iv. If $\bar{\xi}_i > 1/4$ and $\bar{\eta}_i > 1/4$, Player 2 considers the ratio $c := (\bar{\xi}_i - 1/4)/(\bar{\eta}_i - 1/4)$. At stage $i+1$ he then plays a mixed strategy $(q,\, 1-q)$ with $(1-q)/q = c$. Thus if $c > 1$, then Player 1_A is better off than Player 1_B; and Player 2 will play "more" against him, by playing R with a higher probability than L.

It can be shown that any closed approachable set for which the left side of (6.2) does not exceed 1/4 must contain the point $(1/4, 1/4)$; therefore, Player 2 cannot play optimally and force, say, Player 1_A to an average expected payoff lower than 1/4, even if he allows Player 1_B an expected payoff higher than $1/4 = v$.

If p is different from 1/2, the situation is different! Reasoning similar to the above shows that if both players play optimally in $\Gamma(p)$, Player 1_A will get $(1-p)^2$ with probability 1 and Player 1_B will get p^2 with probability 1.

10.7. Conclusions

A start has been made toward a theory of games of incomplete information when often repeated.

It was seen that considerations of long-term revelation or concealment of utilities are crucial for understanding the situation. Specifically, we saw that when one of the two players (Player 1) has special information not available to the other (Player 2), then *he can use this information to his advantage only to the extent that he reveals it.* Whether or not such revelation is worthwhile depends on the circumstances, but at least for the class of games considered here we are able to say when it is worthwhile and when it is not. In the examples we examined in Sections 10.2 and 10.3, one of two extremes of behavior was optimal for Player 1: he must either make no use whatever of his information, or he must disregard the future, playing each stage as if it was the last one (which allows him to make full use of the information). At one time we conjectured that one of these two extremes of behavior is always optimal, but the example of Section 10.4 shows that this conjecture is false. However, it follows from the theorem of Section 10.5 that Player 1 can always play so as to reveal everything he is going to reveal already in the first stage, and then play so as to make no further use of his information.

The most startling conclusion of our investigation is that to understand a situation in which it is known that Player 2's beliefs can be described by a specific subjective probability $p = p_0$ it may be (and usually is) necessary to analyze other situations in which p is different from p_0. Thus, if you know that one of the players in a game places $1/2 - 1/2$ probabilities on a certain conjecture, in order to analyze the situation properly you cannot restrict yourself to the $1/2 - 1/2$ distribution, but *must* consider all other *hypothetical* probability distributions. This surprising conclusion, like the statement italicized in the previous paragraph, is an example of the kind of insight that may eventually have influence on policy-making bodies.

10.8. Where Do We Go From Here?

All the material presented above concerns situations which may be modelled as nature choosing one of a number of 2-person 0-sum games and informing one player, but not the other, of the game that has been chosen; the chosen game is then repeated a large number of times. From here we can (and should) proceed in two main directions:

(a) Extending the theory to a broader class of situations;
(b) Completing and deepening the results in the area of this chapter.

In direction (a), the theory should be extended to 2-person 0-sum situations in which there is incomplete information on *both* sides. Also, non-zero sum two-person games, which are typical of bargaining situations, should be considered. An important difference between such games and 0-sum games is that even in the case of complete information, a repeated non-zero sum game leads to outcomes that are very different from those of the nonrepeated game. Typically, repetition pushes the players towards cooperative behavior, even if the framework is non-cooperative. It remains to be seen what effect incomplete information has on this and other non-zero-sum phenomena. Both cooperative and non-cooperative models should be studied. At a later stage, n-person games with $n > 2$ may be studied.

Averaging methods other than the one used here should be investigated; for example, one might look into discounting methods.

The insights that we hope to gain from the non-zero sum study should be applied also to more general situations, in which possibly *different* games are played successively, all involving one or more parameters that are unknown to one of the sides. For example, consider a situation in which the successive stage games are different, all have payoffs in money, and each player's utility for money is unknown to the other. The question is, to what extent should such utilities be revealed, or concealed, during the course of the "grand" game. This is the question that must eventually be tackled, for it is rarely that the same game is actually repeated again and again: rather, the players engage in different games, but often use the same basic considerations in each of many different situations.

In direction (b), explicit optimal strategies for Player 2 should be sought for all games in the class considered in this chapter; up to now they are known only for special games. As we have seen in Section 10.6, investigation of these optimal strategies has led to the conclusion that many of the games under consideration have a unique value *pair* rather than just a unique value; that is, a value can be assigned to each of the two types A and B of Player 1. For example, when $p = 1/4$ in the game considered in Section 10.6, and both players use optimal strategies, then type A must get 1/16 (with probability 1, not only with expectation 1), and type B must get 9/16. However, such a situation does not always obtain. This phenomenon looks important and should be investigated further.

Other aspects that are not yet understood fully concern the relationship between v_1 and $v_\infty(=\operatorname{cav} u)$. In the example of Section 10.4, we found that Player 1 has an optimal strategy in which he always disregards the future and plays as if he were in a one-stage game. In other words, Player 2 gains nothing from the repetition, in contrast to the example of Section 10.2, in which Player 1 gains nothing from his extra information. The example of Section 10.4 also had the startling property that $v_1 = v_\infty$. There is some reason to believe that this property is related to that of playing each stage as if it was the last one, but we have not succeeded in clarifying this relation (if it exists).

The final problem under heading (b) concerns the second-order properties of the limiting process. For example, consider again the problem of Section 10.5. Rather than considering averages, let us now consider the sum of the payoffs, so that the value of the n-stage game is nv_n. We know that $nv_n = n/4+\mathrm{O}(\sqrt{n})$. Is this error term the best possible? Note that $\pm\sqrt{n}$ is precisely the order of magnitude of the fluctuations that one might expect from probabilistic considerations. This similarity suggests the following approach to answering the above question: In the n-stage game, Player 1 can guarantee $n/4$ by playing $s_i = t_i = 1/2$ at each stage; i.e., by not using his information. In that case, the observed L–R ratio would differ from 1 by $\mathrm{O}(1/\sqrt{n})$. In an optimal strategy, Player 1 would, of course, obtain more by using his information. Thus, he should usually play s_i slightly different from t_i. If this difference is too great too often, then Player 2 would "catch on," and deduce Player 1's type, 1_A or 1_B, with high probability. But if Player 1 keeps the difference between s_i and t_i low enough so that the L–R ratio differs from 1 by $\mathrm{O}(1/\sqrt{n})$, then perhaps he could keep Player 2 from "catching on." This would indicate that $\mathrm{O}(\sqrt{n})$ is indeed the best possible error term. Is this correct?

Postscripts to Chapter Ten

Postscript a is devoted to a discussion of martingales and their relationship to the contents of this chapter. In Postscript b, we discuss a simplified (but less constructive) proof of Lemma 5.3 (that Player 1 can assure himself cav $u(p)$, where cav u is the concavification of u). In Postscript c, the second order properties of the limiting process — i.e., the error term in the formula (5.24) for the value v_n of the n-stage game — are investigated. In Postscript d, explicit optimal strategies for the uninformed player (Player 2) are described. In Postscript e, we show that v_n is a non-increasing function of n; this follows from the so-called "recursive structure" of Γ_n — a formula that expresses the function $v_n(p)$ in terms of the function $v_{n-1}(p)$.

In these postscripts, we denote Players 1 and 2 by P1 and P2, respectively.

a. *Martingales*

Intuitively, underlying this chapter — and indeed, much of the whole book — is the theory of martingales. A *martingale* is defined (Section 10.5, Step 5) (see Doob [1953]) as a sequence of random variables $x_1, x_2, \ldots$, with $\text{Exp}(x_{i+1} \mid x_i) = x_i$ for each i. If P2 knew P1's mixed strategy in Γ_∞, then the sequence of his posteriors p_i that P1 is really of type A would constitute a martingale.[13] The *martingale convergence theorem* [Doob, 1953] asserts that a bounded martingale converges almost surely. It follows that the p_i converge

[13]See (5.6). Indeed, posterior (conditional) probabilities always form a martingale as more and more information becomes available. Explicitly, let E be an event with probability p, and let S be a signal that may be received about the state of the world. Then the conditional probability of E given the signal S is a random variable that depeneds on the true state of the world; the expectation of this random variable is p. It follows that if $S_1, S_2, \ldots$ is a sequence of such signals, and $p_{i+1} := \text{Exp}\,\text{Prob}(E \mid S_1, S_2, \ldots, S_i)$, then the sequence $p_1, p_2, \ldots$ constitutes a martingale.

almost surely, since they are bounded by 1. The limit of the p_i is a random variable, which we call p_∞.

The main theorem of this chapter (Theorem 5.1) has two parts: that P1 can get at least cav $u(p)$ and that P2 can hold the payoff down to cav $u(p)$. To prove the latter, we make use of the principle that P1 cannot use his information without revealing it (see footnote 15 below), and that partial use is associated with partial revelation. Explicitly, if at stage i, P1 uses different strategies s_i and t_i depending on whether he is of type A or B, then the posterior probability p_{i+1} will be different from p_i. Moreover if s_i and t_i are *very* different, then p_{i+1} is likely also to be very different from p_i (unless p_i is close to 0 or 1). From $p_i \to p_\infty$ it follows that p_{i+1} is eventually close to p_i, and hence, unless $p_\infty = 0$ or 1, that s_i and t_i are almost equal; i.e., that P1 plays in a way that is almost independent of his type. He then gets a payoff that is close to the payoff $u(p_i)$ that he could guarantee by playing totally independently of his type, and this, in turn, is close to $u(p_\infty)$. It may be seen that a similar conclusion holds when $p_\infty = 0$ or 1. This reasoning holds for all sufficiently large i, and in Γ_∞, these are what determine the payoff. We thus conclude that the largest expected payoff that P1 can guarantee by any particular mixed strategy is $= \operatorname{Exp} u(p_\infty) \leq$ Exp cav $u(p_\infty) \leq \operatorname{cav} u(\operatorname{Exp} p_\infty) = \operatorname{cav} u(p)$, where the first inequality follows from $u \leq \operatorname{cav} u$, and the second from the concavity of cav u.

Of course, the above is only meant to be suggestive, it is not a proof. The proof is given in the body of this chapter.

b. *Convexity and Concavity as Monotonicity in Information*

The probability of a person P for an event E is a random variable that depends on P's information about the state of the world. Convexity of a function w of P's probabilities p may be interpreted as a kind of monotonicity of w as a function of the information available to P: the more information, the larger, "on the average", is w.

Explicitly, when P gets more information, his probability p for E is replaced by a distribution of posterior probabilities p' whose expectation is p. Convexity says that for all p and all such distributions p',

$$w(p) = w(\operatorname{Exp} p') \leq \operatorname{Exp} w(p'),$$

which is precisely what we mean by w increasing "on the average". A dual statement applies to concavity.

This fundamental fact yields an alternative proof that P1 can get at least $\operatorname{cav} u(p)$ in Γ_∞; i.e., of Lemma 5.3, which says that $\underline{v}_\infty(p) \geq \operatorname{cav} u(p)$, where $\underline{v}_\infty(p)$ is the most that P1 can guarantee in $\Gamma_\infty(p)$. In the body of this chapter, this is proved by explicitly exhibiting a strategy for P1 that guarantees $\operatorname{cav} u(p)$. Alternatively, the result follows if one shows that $\underline{v}_\infty(p) \geq u(p)$ and that $\underline{v}_\infty$ is concave in p. That $\underline{v}_\infty(p) \geq u(p)$ follows from the fact that by ignoring his information completely, P1 can always obtain at least $u(p)$. That $\underline{v}_\infty$ is concave means that P2's corresponding payoff $-\underline{v}_\infty$ is convex, and we have just seen that that simply means that P2's value is monotonic in his information: the more information P2 has, the higher (weakly) is his payoff. In a two-person zero sum game, one can never lose by getting more information, and hence $-\underline{v}_\infty$ is indeed convex. Similar methods show that also $v_n \geq \operatorname{cav} u$.

c. *The Error Term*

The error term in (5.24) has been thoroughly investigated by Zamir [1972] and by Mertens and Zamir [1976, 1977]. Formula (5.24) itself asserts that the error term is always $O(1/\sqrt{n})$. In many cases it is, however, much smaller. For example, in the game of Section 10.2, the error term has order of magnitude $\log n/n$; and in Games 1 and 2 (below), the order of magnitude is $1/n$. Of course, the error term may vanish identically; for example, in the game of Section 10.3, $v_n(p) = \operatorname{cav} u(p) = 0$ for all n. A game for which the order of magnitude of the error is $1/\sqrt{n}$ — the largest possible — is Game 3.

−2	0
0	−1

−1	0
0	−2

Game 1

1	0
−1	−2

−2	−1
0	1

Game 2

3	−1
−3	1

2	−2
−2	2

Game 3

All these assertions hold for any p with $0 < p < 1$.

More generally, Zamir has proved the following results, which for simplicity we state only for the case $\kappa = 2$ (two possible stage games). For the formulation for arbitrary κ, see Zamir [1972].

Theorem N1. *Suppose $u(p)$ is twice differentiable or all p, with a second derivative that is strictly negative and bounded away from zero.*[14] *Then for each p,*

$$v_n(p) = \operatorname{cav} u(p) + \mathrm{O}(\log n/n).$$

Theorem N2. *Suppose that $(\operatorname{cav} u(p) - u(p))/(p(1-p))^{3/2}$ is strictly positive for $0 < p < 1$, and is bounded away from zero. Then for*

[14] A function f is *bounded away from zero* if there is a positive constant δ such that $|f| \geq \delta$ throughout its domain. Such a function need not be bounded.

each p,

$$v_n(p) = \text{cav}\, u(p) + O(1/n).$$

For example, in the game of Section 10.2, we have $u(p) = \text{cav}\, u(p) = p(1-p)$; thus $u''(p) = -2$, which is bounded away from zero. Hence Theorem N1 yields an error term of $O(\log n/n)$. In fact, the error term is asymptotic to $\log n/16n$.

In Game 1,

$$u(p) = -\frac{2}{3} - \frac{p(1-p)}{3};$$

Thus $\text{cav}\, u(p) = -2/3$ identically, so we have $\text{cav}\, u(p) - u(p) = p(1-p)/3$. In Game 2,

$$u(p) = \max\{-p, -(1-p)\};$$

thus $\text{cav}\, u(p) = 0$ identically, and $\text{cav}\, u(p) - u(p) = \min\{p, 1-p\}$. In both these cases, therefore, Theorem N2 yields error terms that are $O(1/n)$.

In Game 3, $u(p) = \text{cav}\, u(p) = 0$ identically for all p. Hence the error term $= v_n(p)$. Mertens and Zamir [1976] have shown that in this example, $\lim \sqrt{n} v_n(p)$ exists and equals the normal density at the p'th quantile x_p of the normal distribution; i.e.,

$$v_n(p) \sim e^{-x_{\rm p}^2/2}/\sqrt{2\pi n}$$

as $n \to \infty$, where $\sim$ means "asymptotic to", and x_p is such that

$$p = \frac{1}{\sqrt{2\pi}} \int_{-\infty}^{x_{\rm p}} e^{-x^2/2}\, dx.$$

Geometrically, the hypothesis of Theorem N1 implies that u is strictly concave and smooth; since cav u is always the concavification of u, this implies that $\text{cav}\, u = u$. The hypothesis of Theorem N2 implies that except at the end points ($p = 0$ and $p = 1$), u lies strictly below cav u (which implies that cav u is linear). At the endpoints, the graphs of cav u and u may meet at positive angles (as in Section 10.3 and in Game 1), or they may be asymptotic; in the latter case, the hypothesis forbids the asymptotic approach from being too close. For example, the conclusion will not in general hold when $\operatorname{cav} u(p) - u(p)$ has the order of magnitude p^2 as $p \to 0$.

Theorems N1 and N2, and Game 3, represent three situations that are conceptually very different as far as P1's long-run policy is concerned. The concavity of u in Theorem N1 implies that for the long run, it is positively harmful for P1 to reveal his type, even partially, and hence he had better use his information very sparingly. The situation in Theorem N2 is the precise opposite: since u lies strictly below its concavification cav u, it is vital for the long run that P1 use his information essentially fully, and essentially from the very beginning, even though he thereby reveals it. Indeed, as long as he does not use his information at all, he will be getting $u(p)$ rather than cav $u(p)$; and making only partial use of his information — revealing partially — is like moving to different probabilities q in the interior of the interval, so that $u(q)$ is still considerably smaller than cav $u(q)$. Game 3 represents a situation that in a sense lies exactly in between these two extremes: both u and cav u are flat, so from the long-run point of view, it does not matter whether P1 reveals his information or not.

Heuristically, these differences explain why it is plausible that the orders of magnitude of the error terms in Theorems N1 and N2 are smaller than in Game 3. In the situations of Theorems N1 and N2, any revelation or concealment will have a significant, first-order effect on the main term cav $u(p)$ in (5.24) (the asymptotic formula for the value v_n of the n-stage game when n is large). Considerations of maximizing the main term will therefore dominate the thinking of P1; he will not take any action that will prejudice the main term of his payoff. But in Game 3, no amount of revelation or concealment will affect the main term, since $u = \operatorname{cav} u$ and the graph of these functions is a straight line. Therefore P1 may orchestrate the pace of revelation so as to maximize the error term, and that is what leads to the relatively large error term in this game.

The above examples illustrate these ideas. Let us start by taking $p = 1/2$ in Game 2. If the game is played only once and P1 is of type A (the left matrix), then it is optimal for him to play top; for type B, on the other hand, bottom is optimal. Moreover, these strategies are strictly dominating (i.e., strictly better for P1 than his other choice, no matter what P2 does). Suppose now that the game is played n times. If P1 makes full use of his information immediately, i.e., plays top or bottom at the first stage, according to whether he is of type A or B, then he reveals his information[15] for all subsequent stages. Thus we are led to a situation in which the true matrix is common knowledge; therefore, starting at the second stage, the payoff will be 0 (the value of both the left and right matrices). But at the first stage itself, P2 does not yet have the information; he cannot do better than randomizing half-half, yielding a payoff of 1/2 to P1 for that stage. Since the payoff is 0 at all subsequent stages, the average payoff is $1/2n$, which is the precise order of magnitude for the error term that is given by Theorem N2. Thus we see that in this case, P1 cannot do better than making use of his information immediately, at least insofar as the order of magnitude is concerned.

The situation in Game 1 is similar, but slightly more subtle. If it were common knowledge that the left matrix is being played, it would be optimal for P1 and P2 to use mixed strategies that assign probabilities 1/3 to "Top" and "Left" respectively; for the right matrix, these probabilities would be 2/3. In this case, though, it is not clear what it means for P1 to "make full use of his information immediately". Indeed, if P1 were to use the above mixed strategy from the very beginning, then his expected payoff at each stage would be exactly $\operatorname{cav} u(p) = -2/3$; i.e., the error term would vanish completely, and thus P1 would not gain even a second order advantage from his superior knowledge (over and above what he

[15]We cannot emphasize too heavily that this "revelation", as well as all the other "revelations" discussed here, are "as if". For P2 actually to deduce anything about P1's type from P1's actions, he would have to know which mixed strategy P1 is using in the n-stage game Γ_n; and of course, he does not in general really know this. What justifies the "as if" argument is the minimax theorem, according to which an optimal mixed strategy for P1 in Γ_n assures him at least the value v_n of the game no matter what P2 does, *even if P2 knows that P1 is using this strategy.* See the discussion just before Step 1 in the proof of Theorem 5.1. In brief, we say that P1 is making a "revelation" each time that his mixed strategy calls for his two types to behave differently.

gets simply by revealing his type to P2). Moreover, he would not be "revealing fully immediately;" since P1's action at the first stage depends on his type only probabilistically, P2 would not be able to deduce P1's type from P1's action with certainty. Assume, for example, that $p = 1/2$ originally. If P1 is really of type A, and the pure choice actually made by his mixed strategy at the first stage is "bottom" (as is more probable), then after the first stage, P2's posterior probability that P1 is indeed of type A is not 1 but 2/3. The fact that the revelation is not full leads one to think that perhaps the *use* that P1 is making of his information also is not as full as it could be.

And, indeed, while P1 cannot get much more than $-2/3$, he *can* do a little better. By considerations of symmetry, P2 will play half-half at the first stage. Suppose now that with certainty, P1 plays bottom if he is of type A, and top if he is of type B. Then he reveals his type fully, but his expected payoff at the first stage is $-1/2$ rather than $-2/3$. He thus gains 1/6 over cav $u(p)$ at the first stage, and nothing thereafter; the average, $1/6n$, is again precisely of the order of magnitude predicted by Theorem N2.

Roughly speaking, what is happening in both these examples, as well as in Theorem N2 in general, is the following: As long as P1 fails to make use of his information, he cannot get more than $u(p)$, which is considerably less than the amount cav $u(p)$ that he could be getting. Therefore it is not worthwhile for P1 to postpone making use of his information; he should do so immediately. At the first stage at which P1 does make use of his information, he may gain a one-time advantage (since P2 does not yet have the information at that stage), and this is what leads to the error term of $O(1/n)$.

In actuality, the story is complicated somewhat by the possibility of revelation that is gradual and/or partial. Assume for simplicity that the graphs of u and cav u actually make positive angles with each other at $p = 0$ and $p = 1$; i.e., that $(\text{cav}\, u(p) - u(p))/p(1 - p)$ is bounded away from zero. These conditions ensure that as long as there is any significant deviation from full revelation — any considerable probability that P2's posterior p_i is near neither 0 nor 1 — then P1's per stage payoff Exp $u(p_i)$ will be considerably below the amount cav $u(p)$ that he could have gotten by revealing fully immediately. Any failure to make full and immediate use of the information available to P1 is thus punished by an amount whose order of magnitude is $1/n$ times the *total number of stages* during which this information is not fully used. On the other hand, each partial revelation is accompanied by a *one-time* reward, which can change the payoff in Γ_n by at most $O(1/n)$. Under these circumstances, it is obviously not worthwhile for P1 to incur any significant delay in making full use of his information. Therefore about the best that he can do is to reveal fully and immediately, and to take his one-time profit from P2's one-time ignorance. We thus again get $O(1/n)$ for the error term.

The situation in Theorem N1 is in a sense opposite to the above. Because u is strictly concave, revelations cause long-term damage; but like above, they may, and usually do, have one-time rewards. If the error term is to be positive, it can only be as the cumulative result of one or more one-time gains; but such gains can only come as the result of revelations, which cause long-term damage that cancels out much of the one-time gains. This explains the relatively small error term in this case.

A detailed analysis is, however, more involved in this case than in Theorem N2. One option of P1 is to reveal nothing until the last stage, and only then to make full use of his information. In this way he will make a one-time gain without incurring any long-term damage. But a single one-time gain can only contribute $O(1/n)$ to the error term. An error term of the size indicated by Theorem N1 can only be achieved by making revelations at many stages.

How worthwhile a single revelation is — if at all — depends on two factors: its timing and its "size". The timing of the revelation — how late it comes in the game — has little or no effect on the one-time gain; but from the long-term point of view, the later it comes, the less the damage. The size of the revelation — how much it causes P2's posterior p_i to shift — affects both the one-time reward and the long-term damage. Let us examine each of these two effects separately.

The one-time gain is usually roughly proportional to the distance $\|s_i - t_i\|$ between the i'th stage mixed strategies s_i and t_i of $P1_A$ and $P1_B$; compare (5.9) and (5.19). This, in turn, is $\mathrm{O}(\varepsilon_i)$, where ε_i measures the size of the revelation, i.e., by how much the posteriors p_i change between Stages i and $i+1$; precisely, ε_i may be defined as $\sqrt{\mathrm{Exp}((p_{i+1} - p_i)^2 \mid p_i)}$, the conditional standard deviation of p_{i+1} given p_i (compare (5.11) and (5.23)). Since the payoff in Γ_n is obtained by averaging over all n stages, we conclude that the one-time gain is of order ε_i/n at most.

The long-term damage, on the other hand, is proportional to the loss in conditional expected payoff, $\mathrm{Exp}((u(p_i) - u(p_{i+1})) \mid p_i)$. By expanding u in a Taylor series around p_i, and noting that the linear term disappears (by the martingale condition $\mathrm{Exp}(p_{i+1} \mid p_i) = p_i$), we see that this $\approx (-u''(p_i)/2)\,\mathrm{Exp}((p_{i+1}-p_i)^2 \mid p_i)$, which is of order ε_i^2. The per-stage loss is therefore of order ε_i^2/n. Since $n-i$ stages remain at the time of the revelation, we conclude that the long-term loss is of the order $(n-i)\varepsilon_i^2/n$.

Summing up, a revelation can be worthwhile only if the order $(n-i)\varepsilon_i^2/n$ of the loss is not larger than the maximal order ε_i/n of the gain. Thus we must have $\varepsilon_i = \mathrm{O}(1/(n-i))$. For the total of all one-time gains we therefore get

$$\mathrm{O}\left[\sum_{i=1}^{n-1} \varepsilon_i\right] = \mathrm{O}\left[\sum_{i=1}^{n-1} \frac{1}{n-i}\right] = \mathrm{O}\left[\sum_{i=1}^{n-1} \frac{1}{i}\right] = \mathrm{O}(\log n),$$

and so the average is $\mathrm{O}(\log n/n)$. This is the gross; what really interests us is the net, obtained from the gross by subtracting the sum of all the long-term damages. But the net cannot be larger than the gross, so that log n/n is the maximum order of magnitude for the net as well.

Intuitively, because u is smooth and strictly concave, the per-stage long-term damage from a revelation behaves like the square of its size (Figure 10.20), whereas the one-time gain is linear in its size. Thus at any given stage, a revelation is likely to be worthwhile if it is sufficiently small. On the other hand, for the long-term damage, the timing is important. Near the beginning of the game, P1 can

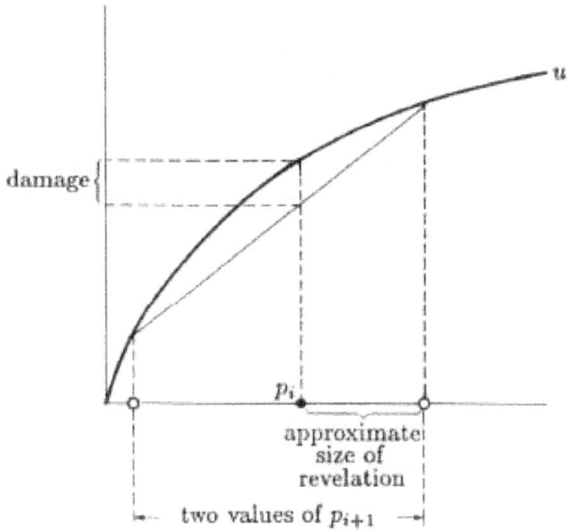

Figure 10.20: The per-stage long-term damage from a revelation.

afford only very small revelations, as he will suffer the long-term damage from them throughout the remainder of the game. As the game progresses, each unit of long-term damage becomes less and less important, and so the revelations can be made larger and larger. Since the gains depend linearly on the sizes of the revelations, we get a series of ever increasing revelations, which leads to the error term in Theorem N1.

We end this note by a brief discussion of the requirements in Theorems N1 and N2 governing the behavior of u near the endpoints of the interval $[0, 1]$. In Theorem N1 the requirement is that u'' be bounded away from 0, and in Theorem N2 it is that u should not approach the linear function cav u too rapidly. The content of both requirements is that the conditions on u that enable the proofs to work be maintained throughout the interval. Thus in Theorem N1, the requirement implies that the order of magnitude of the long-term damage from a single revelation is at least the square of its size, *uniformly* in the posterior p_i; and in Theorem N2, that any failure to reveal immediately causes damage that is "considerable", no matter what p is. If these requirements were not met, then as the posteriors p_i approach the endpoints, one would approach a situation that is conceptually closer and closer to that of Game 3 (where u is linear), and this, like in Game 3, could lead to a larger error term.

Games of the type of Game 3, in which the normal distribution plays a role in the error term, have been investigated further by De Meyer [1989], [1992].

d. *Optimal Strategies of the Uninformed Player*

In Section 10.6, an explicit optimal strategy for P2 was presented, for the specific game treated in Section 10.2. We now show that the same kind of strategy works for any game of the class considered in this Chapter (incomplete information on one side).

Like in the case discussed in Section 10.6, the idea is for P2 to keep track of the payoffs that each of the two possible types of P1 would get. Thus one is led to a game with vector payoffs in the sense of Blackwell [1956]. One then constructs a vector (x, y) such that the area or "quadrant" to the "southwest" of (x, y) is approachable for P2 in an appropriate sense, and such that the expected payoff $px + (1 - p)y$ from this vector is $\operatorname{cav} u(p)$. Any strategy with which P2 can indeed assure approaching this quadrant is optimal for him. Below, we will construct such a strategy explicitly; it is similar to that described in Section 10.6 for the special case treated there.

As sometimes happens, the treatment becomes both notationally simpler and conceptually more transparent in the more general case, when P1 can be any one of κ types, rather than just two. Thus we consider the game with κ-dimensional vector payoffs, each coordinate representing the payoff to a different type of P1. We use x (rather than (x, y)) to denote such a vector payoff, and p (rather than $(p, 1-p)$) to denote the vector of initial probabilities (one for each type). Denote by S_x the "orthant" to the "southwest" of x; that is, the set of vectors ξ each of whose coordinates is $\leq$ the corresponding coordinate of x. Call S_x *e-approachable* ("e-" for expectation) if P2 has a strategy τ in Γ_∞ such that for each strategy σ of P1, the expected average vector payoff for the first n stages of Γ_∞ approaches

S_x as $n \to \infty$. That S_x is e-approachable means that P2 can force the expected vector payoff in Γ_∞ into S_x. The fundamental result is

Theorem N3. *S_x is e-approachable if and only if $q \cdot x \geq \operatorname{cav} u(q)$ for all probability vectors q (not just the* "true" *one p).*

Here $q \cdot x := \sum_{k=1}^{\kappa} q^k x^k$ is the overall (scalar) expectation of the vector payoff x when the probabilities of the coordinates (i.e., of the stage games) are given by q.

Before proving Theorem N3, we show how P2 can use it to construct an optimal strategy in $\Gamma_\infty(p)$. First he must construct a vector x for which

$$S_x \text{ is e-approachable and } p \cdot x = \operatorname{cav} u(p). \tag{N1}$$

Once he has such an x, he can use the explicit strategy described below (at the beginning of the "if" part of Theorem N3's proof) actually to e-approach S_x. His expected payment to P1 is then $\leq p \cdot x = \operatorname{cav} u(p)$; and he can do no better than that, as P1 can guarantee at least $\operatorname{cav} u(p)$ (e.g., by ignoring his information after a possible one-time "split"; see Lemma 5.2). So this is indeed optimal for P2.

To construct an x satisfying (N1), we start by returning to the case of two stage games ($\kappa = 2$). Since $\operatorname{cav} u$ is concave, there is a line $\mathcal{L}$ passing through the point $(p, \operatorname{cav} u(p))$ and supporting the graph of $\operatorname{cav} u$ (Figure 10.21). Define x as the vector whose coordinates are the intercepts of $\mathcal{L}$ with the vertical lines at the endpoints $q = (1, 0)$ and $q = (0, 1)$ of the interval.

To see that this x satisfies (N1), let L be the linear function defined by $L(q) := q \cdot x$. Two points on the line $\mathcal{L}$, namely $((1, 0), x^1)$ and $((0, 1), x^2)$, lie on the graph of L. Since L is linear, it follows that each point of $\mathcal{L}$ is in the graph of L. In particular, the point $(p, \operatorname{cav} u(p))$ is in the graph of L; that is, $p \cdot x = \operatorname{cav} u(p)$. Moreover, $q \cdot x \geq \operatorname{cav} u(q)$ for all q, as $\mathcal{L}$ supports the graph of $\operatorname{cav} u$; hence, by Theorem N3, S_x is e-approachable. Thus both parts of (N1) are verified.

When $\operatorname{cav} u$ is smooth (differentiable) at p, then $\mathcal{L}$ must be tangent to the graph of $\operatorname{cav} u$ at p (Figure 10.21). In that case there

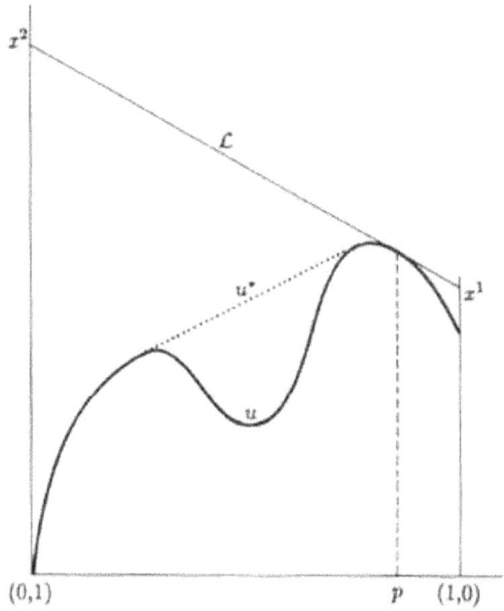

Figure 10.21: The support $\mathcal{L}$ in the smooth case.

is a unique x satisfying (N1), which is easily calculated from p, cav $u(p)$, and the derivative of cav u at p. If cav u has a kink (non-differentiability) at p, then there is a whole range of supporting lines (Figure 10.22). Each of them determines an x satisfying (N1), i.e., an optimal strategy for P2 in $\Gamma_\infty(p)$; and all such x are determined in this way. In the vector payoff space, the set of all such x is an interval.

In the general case, when $\kappa > 2$, the procedure is entirely analogous; one must only replace the line $\mathcal{L}$ by a hyperplane. The intercepts of this hyperplane with the "vertical" lines over the vertices of the q-simplex are then the coordinates of x.

It should be noted that the idea of e-approachability is entirely independent of which probability vector actually represents P2's beliefs. Thus for *every* probability vector p — not just the "true" one — there is an x such that S_x is e-approachable, and $p \cdot x = \operatorname{cav} u(p)$. Consider, for example, the game of Section 10.2. From Theorem N3 it follows that P2 can, if he wishes, play to assure that he will pay no more than 1/9 to $\mathrm{P1}_A$ or 4/9 to $\mathrm{P1}_B$ (Figure 10.23). If the initial probabilities were 2/3 for G_A and 1/3

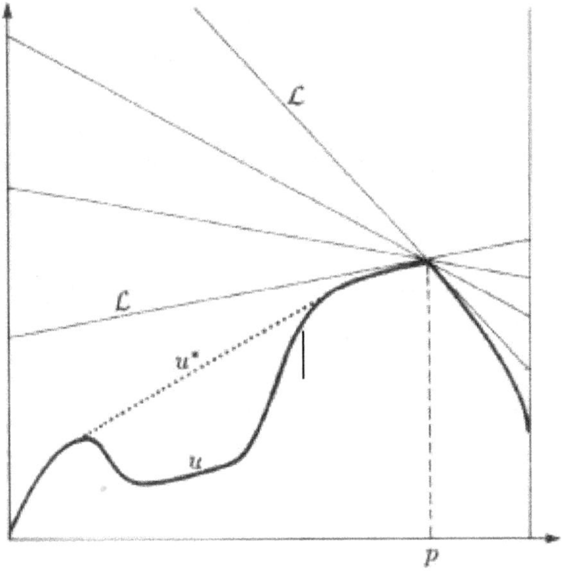

Figure 10.22: Supports $\mathcal{L}$ in the non-smooth case.

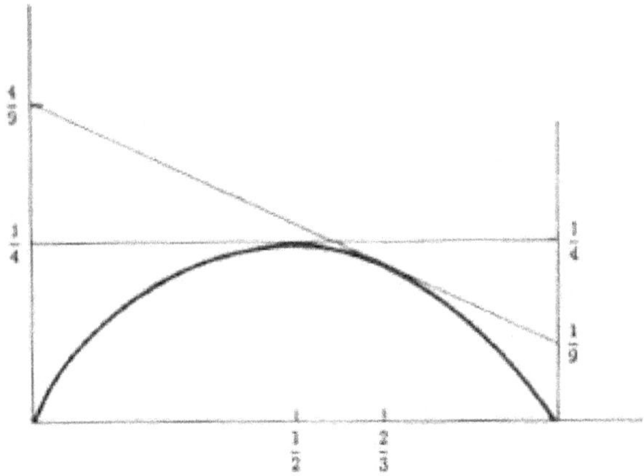

Figure 10.23: Payoff pairs that P2 can assure.

for G_B, this would yield an expectation of 2/9, which is the value $v_\infty(2/3)$ of the infinite game; thus it would be optimal for P2 to play in this way. Since the initial probabilities are, in fact, 1/2–1/2, such play would yield an expectation of 5/18, which is more than the expectation of 1/4 that P2 (who is the minimizer) can assure; so it is not especially worthwhile for him. But if P2 does for some particular reason want to drive $P1_A$'s payoff down to 1/9, he *can* do so — at the expense of paying 4/9 (rather than 1/4) to $P1_B$. In brief, the set of vector payoffs that P2 can assure in Γ_∞ does not at all depend on the initial probability.

We come now to the

Proof of Theorem N3. Only if: By e-approaching S_x, P2 can assure a payoff $\leq q \cdot x$ in $\Gamma_\infty(q)$. But P1 can assure $\geq u(q)$ by ignoring his information and playing optimally at each stage as if the true probability distribution were q. So $q \cdot x \geq u(q)$ for all q. Hence $q \cdot x \geq \operatorname{cav} u(q)$, as $q \cdot x$ is linear in q, and therefore concave.

If:[16] We proceed as in Section 10.6. At the first stage, P2 makes an arbitrary choice. Subsequently, let ξ_n^k be the average payoff to P1's type k up to[17] stage n. Let ζ_n be the closest point to ξ_n in S_x; then at the next stage, P2 plays optimally in $\Delta_1(p_n)$ (the one-shot game in which neither player is informed), where the probability vector p_n is chosen proportional to $\xi_n - \zeta_n$ when $\xi_n \notin S_x$, and is chosen arbitrarily when $\xi_n \in S_x$.

We must show that this strategy assures that the average vector payoff e-appraoches S_x. Assume $x = (0, \ldots, 0)$ (otherwise, subtract x^k from each entry of the k'th stage game G^k). Denote the vector payoff stage n by x_n; let $X_n := x_1 + x_2 + \cdots + x_n$, and define $\hat{X}_n$ by $\hat{X}_n^k := \max(X_n^k, 0)$. Thus $\hat{X}_n$ is in the non-negative orthant of κ-space, and indeed is the closest point in that orthant to X_n. Setting $\hat{\xi}_n := \hat{X}_n/n$, note that $\hat{\xi}_n = \xi_n - \zeta_n$ (Figure 10.24). We claim that

$$\hat{X}_n \cdot \operatorname{Exp}(x_{n+1} \mid \hat{X}_n) \leq 0 \tag{N2}$$

for all values of $\hat{X}_n$. Indeed, $\hat{X}_n = n\hat{\xi}_n = n(\xi_n - \zeta_n) = \alpha_n p_n$, where α_n is a non-negative scalar. So $\hat{X}_n \cdot \operatorname{Exp}(x_{n+1} \mid \hat{X}_n) = \alpha_n \operatorname{Exp}(p_n \cdot x_{n+1} \mid \hat{X}_n)$. But $\operatorname{Exp}(p_n \cdot x_{n+1} \mid \hat{X}_n)$ is just the expectation of the outcome

[16]See the end of the proof for an intuitive sketch.

[17]"Up to" will always mean "up to and including".

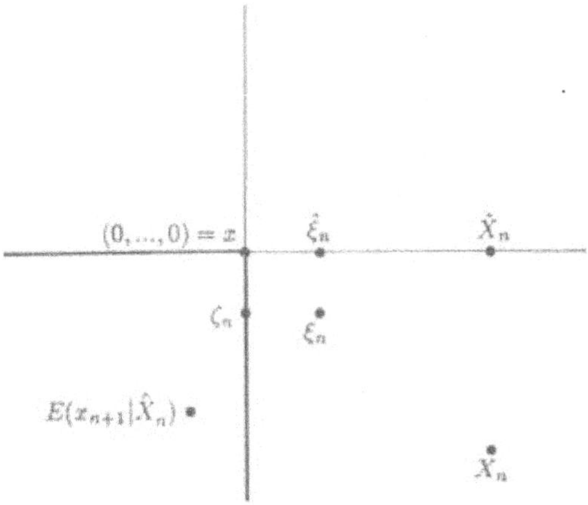

Figure 10.24: Proof of Theorem N3.

when P2 plays optimally in $\Delta_1(p_n)$, so it is $\le u(p_n) \le \operatorname{cav} u(p_n) \le p_n \cdot x = 0$ (since $x = (0, \ldots, 0)$); so (N2) is proved.

Let M be a bound on the absolute values of the entries in the payoff matrices of all the stage games G^k. We claim that

$$\operatorname{Exp} \|\hat{X}_n\|^2 \le 3\kappa M^2 n, \tag{N3}$$

where $\| \ \|$ is the euclidean norm. Proceeding by induction, assume (N3) for n. Set $\hat{x}_n := \hat{X}_n - \hat{X}_{n-1}$. Then $|\hat{x}_n^k| \le |x_n^k|$ for all k, whence $\|\hat{x}_n\| \le \|x_n\| \le M\sqrt{\kappa}$; so

$$\begin{aligned} \operatorname{Exp}\|\hat{X}_{n+1}\|^2 &= \operatorname{Exp}(\|\hat{X}_n + \hat{x}_{n+1}\|^2) \\ &= \operatorname{Exp}(\|\hat{X}_n\|^2 + \|\hat{x}_{n+1}\|^2 + 2\hat{X}_n \cdot \hat{x}_{n+1}) \\ &\le \kappa M^2(3n+1) + 2\operatorname{Exp}(\hat{X}_n \cdot \operatorname{Exp}(\hat{x}_{n+1} \mid \hat{X}_n)). \end{aligned}$$

Thus it suffices to show that $\hat{X}_n \cdot \operatorname{Exp}(\hat{x}_{n+1} \mid \hat{X}_n) \le \kappa M^2$ for all values of $\hat{X}_n$; so by (N2), that

$$\hat{X}_n \cdot \operatorname{Exp}((\hat{x}_{n+1} - x_{n+1}) \mid \hat{X}_n) \le \kappa M^2. \tag{N4}$$

Now if $\hat{X}_n^k \ge M$, then *a fortiori* $\hat{X}_n^k \ge 0$, so $X_n^k = \hat{X}_n^k \ge M$, so $X_{n+1}^k = X_n^k + x_{n+1}^k \ge M - M = 0$, so $\hat{X}_{n+1}^k = X_{n+1}^k$, so

$\hat{x}^k_{n+1} = x^k_{n+1}$, so

$$\hat{X}^k_n \mathrm{Exp}((\hat{x}^k_{n+1} - x^k_{n+1}) \mid \hat{X}_n) = 0 < M^2; \tag{N5}$$

and if $\hat{X}^k_n < M$, then $\hat{X}^k_n \,\mathrm{Exp}((\hat{x}^k_{n+1} - x^k_{n+1}) \mid \hat{X}_n) \leq M \max |x^k_{n+1}| \leq M^2$, since x^k_{n+1} and $\hat{x}^k_{n+1}$ have the same sign. Combining this with (N5) and summing over k, we obtain (N4), and so (N3) for all n.

Let δ_n denote the euclidean distance between $\mathrm{Exp}\,\xi_n$ and S_x. Since $\zeta_n \in S_x$ and S_x is convex, it follows that $\mathrm{Exp}\,\zeta_n \in S_x$. So

$$\delta_n \leq \|(\mathrm{Exp}\,\xi_n) - (\mathrm{Exp}\,\zeta_n)\| = \|\mathrm{Exp}(\xi_n - \zeta_n)\| = \|\mathrm{Exp}\,\hat{\xi}_n\| \leq \mathrm{Exp}\|\hat{\xi}_n\|,$$

by the triangle inequality. By Schwarz's inequality and (N3),

$$(\mathrm{Exp}\|\hat{\xi}_n\|)^2 \leq \mathrm{Exp}\|\hat{\xi}_n\|^2 = \mathrm{Exp}\|\hat{X}_n\|^2/n^2 \leq 3\kappa M^2/n; \tag{N6}$$

so we conclude that $\delta_n \leq C/\sqrt{n}$ where C is the constant $M\sqrt{3\kappa}$. This proves Theorem N3.

The idea of the proof is that whenever the average vector payoff ξ_n is outside S_x, P2 drives it towards S_x; i.e., towards the point ζ_n in S_x that is nearest to ξ_n (Figure 10.24). Thus if p_n is a probability vector with direction $\hat{\xi}_n := \xi_n - \zeta_n$, then P2 uses a stage strategy that minimizes the component of x_{n+1} (the next stage vector payoff) in the direction p_n. Since we have taken $x = (0, \ldots, 0)$, the hypothesis of the theorem yields $u(p_n) \leq p_n \cdot (0, \ldots, 0) = 0$; so by playing optimally in $\Delta_1(p_n)$, P2 can actually force a non-positive payoff in $\Delta_1(p_n)$. But the payoff in $\Delta_1(p_n)$ is merely the vector payoff weighted by the p^k_n, i.e., the component of x_{n+1} in the direction p_n. So P2 can force this component to be non-positive. Thus the sum of payoffs does not increase in the direction of p_n, so the average decreases.

The technical expression of this idea is (N2), which says that x_{n+1} is never positively correlated with $\hat{X}_n$. If we could replace[18] $\hat{X}_n$ by X_n we could deduce, by induction, that the variance of the sum X_n of the previous x_i is $\leq$ the sum of the variances of those x_i, and so is $\mathrm{O}(n)$; so $\mathrm{Var}\,\xi_n$ would be $\mathrm{O}(n)/n^2 = \mathrm{O}(1/n)$. In fact, we cannot replace $\hat{X}_n$ by X_n; but by doing so, and then estimating the

[18] Recall that $\xi_n = X_n/n$ and $\hat{\xi}_n = \hat{X}_n/n$.

resulting error ((N4)), we *can* conclude ((N3)) that Var $\hat{X}_n = \mathrm{O}(n)$, which suffices.

Note that Theorem N3 provides an independent proof for that part of the main theorem of this chapter according to which P2 can hold the expected payoff down to cav $u(p)$. One can also go in the opposite direction: derive Theorem N3 from the main theorem of this chapter. In fact, Blackwell's [1956] characterization of convex closed approachable sets, and the relations between approachability and excludability proved by him, can all be derived from the main theorem of this chapter. But for an explicit optimal strategy for P2, we must still use the method of Blackwell described above.

The proof has several other implications worth noting. First, in the course of proving e-approachability, we actually prove something stronger, namely Exp$\|\hat{\xi}_n\| \to 0$ (see (N6)). That is, not only does the expectation of the average vector payoff ξ_n approach the corner S_x, but also the expectation of the distance between ξ_n and S_x tends to 0. It follows immediately that ξ_n approaches S_x in probability, i.e., that for each positive ε there is an N such that for all $n > N$, the probability is at least $1 - \varepsilon$ that ξ_n is within ε of S_x. Using Exp$\|\hat{\xi}_n\| = \mathrm{O}(1/\sqrt{n})$ (see (N6)), and mimicking the proof of the strong law of large numbers, one can establish the still stronger result that ξ_n approaches S_x with probability 1. This, indeed, is Blackwell's [1956] definition of approachability.

The second implication is the speed of convergence, $\mathrm{O}(1/\sqrt{n})$. For one thing, this yields an alternative proof for something already proved in the body of the text, namely $v_n - v_\infty = \mathrm{O}(1/\sqrt{n})$. But it goes further than that. That $v_n - v_\infty = \mathrm{O}(1/\sqrt{n})$ means that P2 can hold P1 to within $\mathrm{O}(1/\sqrt{n})$ of v_∞ in the game Γ_n; but in principle, this might require a strategy for P2 that depends on n from the very beginning. The proof of Theorem N3 implies that P2 has a single strategy in Γ_∞ that assures him an expected payoff $\leq v_\infty + \mathrm{O}(1/\sqrt{n})$ simultaneously[19] in all the games Γ_n.

[19]No such result holds for P1. For example, when u is strictly concave (as in Section 10.2), P1 cannot guarantee any error term at all for all n simultaneously, as even the slightest one-time "revelation" has a permanent first-order effect.

e. *Monotonicity of v_n and the Recursive Structure of Γ_n*

It is relatively easy to prove that

$$nv_n + mv_m \geq (n+m)v_{n+m} \tag{N7}$$

for all n and m, and from this it follows that

$$v_n \geq \lim v_n \tag{N8}$$

for all n. Indeed, (N7) follows from the fact that P2 can play the first n stages in Γ_{n+m} as if he were playing Γ_n, then ignore everything he learned in those stages, and play the last m stages as if he were playing Γ_m. From (N7) we obtain

$$v_n \geq v_{nk} \tag{N9}$$

for all n and k, and then (N8) follows if we let $k \to \infty$.

The above derivation of (N8) uses the fact that $\lim v_n$ exists. This, in turn, is a consequence of the main theorem of this chapter, Theorem 5.1, whose proof lies comparatively deep. One can, however, use (N7) directly to give a comparatively simple proof that $\lim v_n$ exists; from this it follows as in Step 7 in the proof of Theorem 5.1 that also v_∞ (the value of the infinite stage game) exists.

To prove that $\lim v_n$ exists, assume without loss of generality that all the entries in the game matrices G_A and G_B are nonnegative; let M be a bound for these entries. Then the sequence $\{v_n\}$ is also bounded by M; hence lim inf v_n exists and is finite. Let $\{n_j\}$ be a sequence such that

$$v_{n_j} \to \liminf v_n \tag{N10}$$

as $j \to \infty$. Now for an arbitrary n and a fixed j, write $n = qn_j + r$, where $0 \leq r < n_j$. From (N7) and (N9) it then follows that

$$nv_n \leq qn_j v_{n_j} + rv_r \leq nv_{n_j} + n_j M. \tag{N11}$$

Keeping j fixed, divide (N11) by n, and let $n \to \infty$; it follows that $\limsup v_n \leq v_{n_j}$ for all j. Now let $j \to \infty$ and use (N10); it follows that $\lim\sup v_n \leq \lim\inf v_n$, which means that $\lim v_n$ exists, as asserted.

In view of (N8), one may ask whether v_n is monotonically non-increasing as a function of n, ie., whether

$$v_{n+1} \leq v_n \tag{N12}$$

for all n. The answer is "yes", but the proof lies deeper than the above. It was first proved by Stearns in the late sixties, but that proof was never published, even in report form. Given an optimal strategy for P1 in Γ_{n+1}, Stearns constructs a strategy for him in Γ_n that yields him at least as much. Here we present an entirely different proof of (N12), due to Sorin [1979], which is based on the "recursive structure" of Γ_n. This structure is worth exhibiting in its own right, as it plays an important role in many areas of the theory.

Theorem N4. *For any probability vector $p = (p^1, p^2, \ldots, p^\kappa)$, and any n, we have*[20]

$$(n+1)v_{n+1}(p) = \max_\sigma \min_\tau [\gamma_1(\sigma,\tau) + \mathrm{Exp}\, n v_n(p_2)], \tag{N13}$$

where σ and τ range over the strategies of P1 and P2 respectively in Γ_1 (the one-stage game), γ_1 is the payoff function in Γ_1, and p_2 is P2's conditional probability vector after the single stage of Γ_1 has been played (but before he knows his payoff), given that P1's mixed strategy in Γ_1 was indeed σ, and given Pl's pure choice.

As usual, p_2 is a random variable, depending on the pure choice made by P1 in accordance with his mixed strategy σ.

For a proof of (N13), see Zamir [1972]. Intuitively, we may suppose that P1 and P2 use optimal mixed strategies σ^* and τ^* in Γ_{n+1}, which induce mixed strategies σ and τ for the first stage. The expected payoff for that stage is then $\gamma_1(\sigma,\tau)$. Since σ^* is optimal, P1 may announce it without losing anything; and then after the first stage, P2, having observed the pure choice of P1 at the first stage, will have the probability vector p_2 for the true game being played.

[20] The general formulation presented here is due to Zamir [1972].

After the first stage, the players are faced with an n-stage game; and since $\sigma^*$and τ^* remain optimal after the first stage, they must yield an average payoff of $v_n(p_2)$ for those stages, or a total payoff of $nv_n(p_2)$. Thus the total expected payoff for all $n+1$ stages is the expression inside the square brackets on the right side of (N13). The max min of this expression is therefore the value $(n+1)v_{n+1}(p)$ of $(n+1)\,\Gamma_{n+1}(p)$ (the $(n+1)$-stage game with total rather than average payoff), which is what (N13) asserts.

Sorin's proof of (N12) proceeds by induction. Suppose it has been proved up to but not including n. Then by (N13), the concavity of v_n, the induction hypothesis, the martingale condition $\operatorname{Exp} p_2 = p$ (see (5.8) or Postscript a), and again (N13), we have

$$\begin{aligned}
(n+1)v_{n+1}(p) &= \max_{\sigma}\min_{\tau}[\gamma_1(\sigma,\tau) + \operatorname{Exp} nv_n(p_2)] \\
&= \max_{\sigma}\min_{\tau}[\gamma_1(\sigma,\tau) + \operatorname{Exp} v_n(p_2) \\
&\qquad + \operatorname{Exp}(n-1)v_n(p_2)] \\
&\leq \max_{\sigma}\min_{\tau}[\gamma_1(\sigma,\tau) + v_n(\operatorname{Exp} p_2) \\
&\qquad + \operatorname{Exp}(n-1)v_{n-1}(p_2)] \\
&= \max_{\sigma}\min_{\tau}[\gamma_1(\sigma,\tau) + v_n(p) + \operatorname{Exp}(n-1)v_{n-1}(p_2)] \\
&= v_n(p) + \max_{\sigma}\min_{\tau}[\gamma_1(\sigma,\tau) + \operatorname{Exp}(n-1)v_{n-1}(p_2)] \\
&= v_n(p) + nv_n(p) = (n+1)v_n(p),
\end{aligned}$$

and dividing by $n+1$ yields (N12). (The removal of $v_n(p)$ from the max min on the next-to-last line is permitted because $v_n(p)$, unlike $\operatorname{Exp} v_n(p_2)$, does not depend on σ and τ.)

One may begin the induction in the same way; alternatively, one can note that $v_2 \leq v_1$ follows from (N9). This proves (N12).

References

Aumann, R. J. (1964). Mixed and behavior strategies in infinite extensive games, in *Advances in Game Theory, Annals of Mathematics Study* 52, eds. M. Dresher, L. S. Shapley and A. W. Tucker. Princeton: Princeton University Press, 627–650.

Blackwell, D. (1956). An analogue of the minimax theorem for vector payoffs, *Pacific Journal of Mathematics*, **6**, 1–8.

De Meyer, B. (1989). Repeated games and multidimensional normal distributions, CORE Discussion Paper 8932, Université Catholique de Louvain, Louvain-la-Neuve.

De Meyer, B. (1993). Repeated games and the central limit theorem, CORE Discussion Paper 9303, Université Catholique de Louvain, Louvain-la-Neuve.

Doob, J. L. (1953). *Stochastic Processes*, New York: John Wiley & Sons.

Harsanyi, J. C. (1967-8). Games with incomplete information played by Bayesian players, Parts I, II, III, *Management Science*, **14**, 159–182, 320–334, 486–502.

Kuhn, H. W. (1953). Extensive games and the problem of information, in *Contributions to the Theory of Games II*, *Annals of Mathematics Study* 28, eds. H. W. Kuhn and A. W. Tucker, Princeton: Princeton University Press, 193–216.

Mertens, J.-F., and S. Zamir (1976). The normal distribution and repeated games, *International Journal of Game Theory*, **5**, 187–197.

Mertens, J.-F., and S. Zamir (1977). The maximal variation of a bounded martingale, *Israel Journal of Mathematics*, **27**, 252–276.

Sorin, S. (1979). A note on the value of zero-sum sequential repeated games with incomplete information, *International Journal of Game Theory*, **8**, 217–223.

Zamir, S. (1972). On the relation between finitely and infinitely repeated games with incomplete information, *International Journal of Game Theory*, **1**, 179–198.

Index

A

acceptable points, 7–8, 37–38
agreements
  enforceability of, 41–42
  formal, 40–42
  informal, 40–41
  posteriori equilibria, 97
  subjective probabilities, 172
algebra of combining assumption, 47
allocations, 56, 61, 117, 125
Anscombe, F. J., 104
approachability theory, 17–20
arithmetic mean, 13
assumptions, 45–49, 62–63
asymptotic value, 121, 125, 137
Aumann, R. J., 103–104, 116, 118, 181, 198
axioms systems, 83
  of utility theory, 46

B

Babylonian Talmud, 220, 222
bankruptcy problems
  coalition formation, 232
  consistency, 224
  definition of, 224
  discussion of, 220
  game-theoretic analysis, 221
  less-than-half case, 231
  nucleolus and kernel, 232
  proportional division principle, 220
  Talmudic literature, 220, 222, 228, 231, 240
bargaining problem of Nash, 147
bargaining theory, 227, 240
Bayesian decision theory, 193, 198
Bayesian rationality, 189
  common prior assumption, 192–196
  correlated equilibrium, 178–183
  discussion of, 185–192
  game-theoretic viewpoint, 175
  ignorance expression, 196–198
  literature, 198–199
  n-person game, 183–185
  probabilistic nature, 176
  strategy randomization, 196–198
  subjective probabilities, 177
believability concept, 208–209
Bernheim, B. D., 198
Blackwell, D., 17–18, 278, 301

C

cardinal market, 146–149
cardinal utility, 145, 150
cardinal value allocation, 146–147, 149
Cauchy-Schwarz inequality, 272, 275
Champsaur, P., 150
chances, 43–45

classical game theory, 81
coalition formation, 232, 234
commitment, 100, 102–103, 191
commodity, 128, 130, 132
commodity bundle, 56, 60
common knowledge, 169–170, 187–188, 191, 196–197
  definition of, 171
common prior assumption (CPA), 192–193
communication, 99–100, 102
competitive allocation concept, 147
competitive equilibrium, 56–57, 62, 67, 116, 118, 125
compound lottery, 45, 47
concavification, 265, 283, 287–288
conditional probability, 258–259, 268, 271, 274
consistent case, 105, 109, 195
constant partition, 4, 25
constrained equal award (CEA), 228
continuous games, 119
continuum
  of commodities, 59, 61
  of players, 58, 60
  of traders, 116
contract curve, 57, 68
convexity and concavity, 284
cooperative behavior, 202, 281
cooperative game theory, 5–7, 74, 99–101, 103, 202
core, 56–58, 61–62, 68
correlated equilibrium, 178–183, 191
correlated strategy vectors, 3
correlation, 73, 86–87, 99–100, 179
cost-to-punishment ratio, 218
criticism, equilibrium, 177, 206
cross-hatched region, 205

D

Debreu, G., 116, 145
decision-making process, 111
de Finetti phenomenon, 141
DeGroot, M. H., 172
Delphi technique, 172
descriptive modelling, 201
differentiability, 116, 121–123, 137–138
direct equilibrium arguments, 103

E

e-approachable (e-expectation), 294–296
economic community, 193
economic equilibrium, 55
efficiency prices, 134
endogeneity, 190
enforceability of agreements, 40–42
epsilon-delta phraseology, 162
equal division principle, 224
equilibrium allocations, 57–60, 62, 64, 68–70
equilibrium arguments, 103
equilibrium concept of Nash, 176–178, 180, 192
equilibrium payoffs, 85, 88, 96–97
equilibrium points, 83, 86, 89
equilibrium theory, 127
equity, 126, 129
equivalence principle, 115–116
ethical considerations, 130
Euclidean distance, 300
Euclidean space, 60, 121, 123
excludability theory, 17–20
exogeneity, 190
expected payoff, 13–14, 38–40
exponents, 5
extensive game theory, 105
external enforcement mechanism, 101

F

Fair Representation, 240
feasible payoff, 85, 88, 97, 204
Fenchel, W., 137, 139, 156
de Finetti, B., 137
finite markets, 57, 117, 148, 249
finiteness, 188–189
food transfers, 128–129
formal agreements, 40–42
freedom of choice, 185

G

game theory, 99, 104, 192, 279
 analysis, 221
 Babylonian Talmud, 222
 bankruptcy problems, 221
 cooperative notions, 202
 game description, 82
 viewpoint, 175
Gaon, Rabbi Hai, 224
Gaussian curvature, 122, 138–139, 158
gradual disarmament procedure
 behavioral strategies, 257
 concavification, 265
 mathematical model, 243
 optimal strategy, 255–256, 261, 263
 subsequent stage, 261
 utility theory, 243
guaranteed value arguments, 103

H

Hamilton, Alexander, 240
Harsanyi doctrine, 108–109, 172, 183
Harsanyi, John, 78, 105–106, 108, 171, 195–196, 246, 248
homogeneity, 149
horse lotteries, 45–50
hypothetical probability distributions, 280

I

identity partition, 4, 12
impersonal tastes, 193
impossible/negligible concept, 97
inconsistent case, 105, 109, 195
indirect equilibrium arguments, 103
individual trader
 marginal contribution, 117
 non-atomic case, 124
 non-infinitesimal bundle, 62
informal agreement (gentlemen's agreement), 40–41
information system, 189–190
information vector, 11–12, 18
inherently unstable games, 37
initial allocation, 123, 127, 132
interpretation, 126–130

J

Jerusalem Talmud, 232

K

Kadane, J. B., 198
Kahneman, D., 172
Kannai, D., 120, 137
Kannai, Y., 225
Kleitman, D., 225
Krawczyk, J., 218

L

Lagrange multiplier, 158
Larkey, P. D., 198
less-than-half case, 231
logical probability, 43
long term competition
 cooperative behavior, 202
 descriptive modelling, 201
 payoff limiting average form, 212–218
 PEPs characterization, 209–212
 perfect equilibrium points, 207–209
 repeated game models, 203–207
 solution concept, 201–202
long-term damage, 291–293
lotteries, 44–45
Lyapunov's theorem, 91, 93

M

marginal contribution, 117, 160
market model, 118, 123–126, 144
markets values
 allocation, 117
 cardinal context, 144–149
 continuous games, 119–121
 differentiability, 121–123
 equivalence principle, 116
 example of, 130–132

finite games, 118–119
interpretation, 126–130
literature discussion, 149
marginal contribution, 117
market equilibrium, 56
market model, 118, 123–126
non-atomic market, 116
results statement, 123–126
Shapley value, 115
smoothness, 121–123, 137–143
social welfare functions, 117
uniformity conditions, 143–144
unrigorous demonstration, 132–137
value allocations, 116, 118
markets with continuum of traders, 58–59
martingale convergence theorem, 271, 283
Maschler, M., 75, 103
Mas-Collel, A., 137
matching pennies, 75, 81, 196
mathematical model, 243–244
Mertens, J.-F., 188, 285, 287
minimax theorem, 90, 272
min-max payoff, 204–205
Mishna, 220–225
mixed strategy equilibrium point, 79, 190–191, 196
mixing value, 125
modern economic theory, 194
monotonic linear transformation, 85
Morgenstern, O., 103, 111, 194

N

Nash classical theory, 87, 106
Nash equilibrium payoffs, 78–80, 87, 95
Nathan, Rabbi, 222
n-dimensional vector, 163
Neumann, J. von, 103, 111, 194
non-atomic games theory, 133
non-atomic market, 116, 123
non-cooperative game, 100, 202
non-cooperative n-person theory, 74
non-cooperative theory, 99, 103, 281
non-transferable utility (NTU) game, 148–149
normal distribution, 287
notation, 2–5
n-person games, 1, 73
acceptable points in, 7–8, 37–38
approachability and excludability theory, 17–20
cooperative, 5–7
notations, 2–5
payoffs in, 10–16
strong equilibrium points, 16–17
theorem, 20–36
two-person, zero-sum games, 8–10
n-tuple of strategies, 83, 85
null player, 119

O

objective, 73, 85, 104
objective mixed equilibrium payoff, 85
one-time reward, 291
optimal strategy, 255–256, 261, 263, 276
ordinal markets, 146–148
outside observer perspective, 186

P

Pareto optimal allocation, 68, 147
payoffs
actual, 14
average payoff, 272
common unit of measurement, 6
definition of, 104–105
distribution, 4, 127
expected, 13–14, 38–40
function, 85
limiting average form, 212–218
Nash equilibrium, 181
in supergame, 10–16
payoff vector, 118, 204–205, 235
Pearce, D. G., 198
pegging strategies, 89
Peleg, B., 75
perfect competition, 55–57

perfect equilibrium points (PEPs), 207–209
believability concept, 209
characterization of, 209–212
cost-to-punishment ratio, 218
damage/cost ratio, 217
perfectly competitive market, 116
physical probability, 43, 193
posteriori equilibria, 97–98
posteriori preferences, 98–99
posteriori theory, 99
practical method, 172
preference relation, 77, 123, 140, 146
preference scale, 145–150
price vector, 62, 125
principle of insufficient reason, 111
prisoner's dilemma, 204–205
probability (*see also* subjective probability), 43, 85
definition of, 104
proportional division principle, 220
punishment, 202, 207, 210
pure exchange economies, 56

Q

quasi-concave utilities, 122, 137–138, 141

R

Radon-Nikodym derivatives, 189
random device, 3, 73, 80, 190
randomized strategy
definition of, 83
formal model, 82
randomizing structure, 88
rationalizability theory, 198
reduction of compound lotteries assumption, 47
repeated game models, 203–207
Rosenthal, R. W., 75
roulette lotteries, 45–46

S

Savage, L. J., 50–52, 109, 177, 193
self-consistent, 227
self-duality, 228, 230
self-enforcing agreement, 97, 101, 117
Selten, R., 78, 106, 208
shadow prices, 134
Shapley, L., 118
Shapley value, 115, 118, 129–130, 202
side payments, 6, 68, 127
simple lotteries, 45
single game-theoretic tool, 176
Small Worlds theme, 109
smoothness, 121–123, 137–143
smooth preferences, 116
social welfare functions, 117, 126
solution concepts, 201–202, 235
stable games, 38
strong equilibrium points, 16–17, 23, 40
subjective correlated equilibrium, 194–195
subjective probability theory, 73, 75, 85, 104, 110
assumption of utility theory, 45–47
chances, 43–44
equilibrium payoff vectors, 74
literature, 50–52
lotteries, 44–50
n-person games, 74
probability theory, 73
random devices, 73
random variable, 74
two-person zero-sum games lose, 73–75
substitutability assumption, 47, 119
supergame c-strategy vector, 1, 7, 11, 13
payoff to, 13
strategies and payoffs in, 10–16
summable, 15–16

T

Talmudic literature, 228, 231
Taylor series, 292
tentative agreement, 209

three-person, zero-sum games, 37, 76–79
trade, 56
traders
  commodity, 130
  infinitesimal subset, 124
  marginal contribution, 160
  trader space, 123
  true contributions, 118
  utility function, 127
traditional cooperative outcome, 205
transferable utility (TU), 127–128, 131–132, 149
transferable utility competitive equilibrium (TUCE), 151
true contribution, 160–161
true correlation, 179
Tucker, A. W., 10
Tversky, A., 172
two-dimensional vectors, 278
two-person, zero-sum game, 8–10, 73–78, 81, 90, 176, 181
  equilibrium points, 89
  utility functions, 90
  value concept, 73

U

uniform smoothness, 116, 124
uninformed player, 294
universality, 191–192
utility functions, 99, 124, 145
utility theory, 60, 243
  application of, 53
  assumptions, 45–47
  axioms of, 46
  roulette lotteries, 61
  von Neumann and Morgenstern, 44–45, 111

V

value allocations, 116, 118, 125, 137
  concept of, 145
  definition of, 125, 145
  food transfers, 128–129
  utility function, 127
Value Equivalence Theorem, 116–118, 125, 127, 137, 148
vector function, 4, 11, 60
vectors of probabilities, 104
von Neumann–Morgenstern utility theory, 44–45
von Neumann's minimax theorem, 205

W

Walrasian markets, 150
Walrasian pure exchange economies, 116
Washington, George, 240

Z

Zamir, S., 188, 285–287, 303

www.ingramcontent.com/pod-product-compliance
Ingram Content Group UK Ltd.
Pitfield, Milton Keynes, MK11 3LW, UK
UKHW021843090625
459472UK00006B/36